5th edition

# geog.1

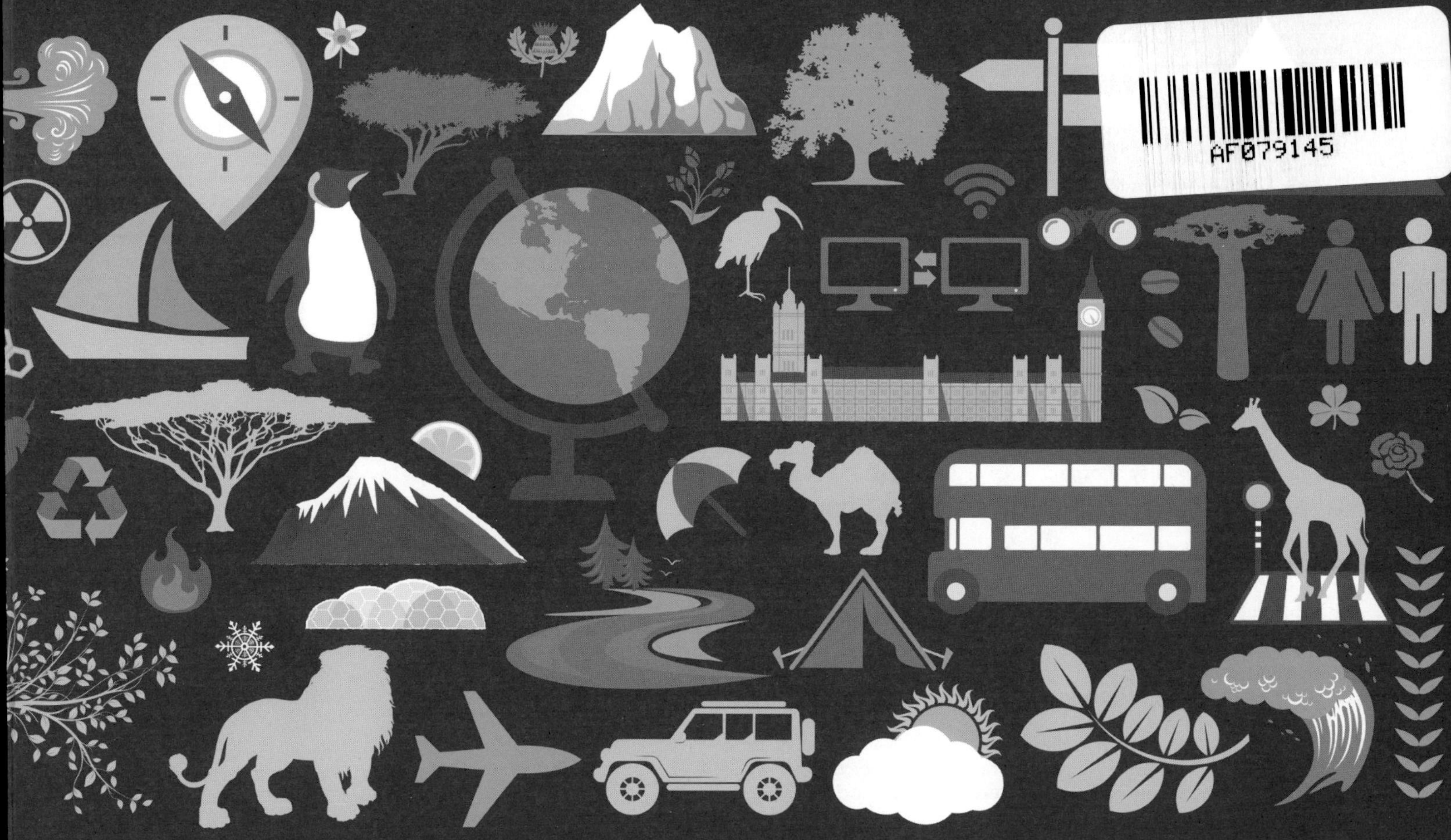



## workbook

<justin woolliscroft>

Name: ..................................................

Class: ..................................................

OXFORD

Great Clarendon Street, Oxford, OX2 6DP, United Kingdom

Oxford University Press is a department of the University of Oxford. It furthers the University's objective of excellence in research, scholarship, and education by publishing worldwide. Oxford is a registered trade mark of Oxford University Press in the UK and in certain other countries

Author: Justin Woolliscroft

The moral rights of the authors have been asserted

Database right Oxford University Press (maker)

First published 2006

New editions published 2008 and 2014

This edition published 2019

All rights reserved. No part of this publication may be reproduced, stored in a retrieval system, or transmitted, in any form or by any means, without the prior permission in writing of Oxford University Press, or as expressly permitted by law, by licence or under terms agreed with the appropriate reprographics rights organization. Enquiries concerning reproduction outside the scope of the above should be sent to the Rights Department, Oxford University Press, at the address above.

You must not circulate this work in any other form and you must impose this same condition on any acquirer

British Library Cataloguing in Publication Data
Data available

ISBN: 978-0-19-844606-4

16

Paper used in the production of this book is a natural, recyclable product made from wood grown in sustainable forests.
The manufacturing process conforms to the environmental regulations of the country of origin.

Printed and bound by CPI Group (UK) Ltd, Croydon, CR0 4YY

## Acknowledgements

The publisher would like to thank the following for permissions to use photographs and other copyright material:

**Cover:** Shutterstock/OUP
**p26:** OUP; **p28:** © Tony Waltham; **p32:** Andrew Leaney

Ordnance Survey (OS) is the national mapping agency for Great Britain and a world-leading geospatial data and technology organisation. As a reliable partner to government, business and citizens across Britain and the world, OS helps its customers in virtually all sectors improve quality of life.

Some pages or activities in this edition are based on material written by Anna King, Jack Mayhew, and Susan Mayhew for the original edition.

Links to third party websites are provided by Oxford in good faith and for information only. Oxford disclaims any responsibility for the materials contained in any third party website referenced in this work.

Every effort has been made to contact copyright holders of material reproduced in this book. Any omissions will be rectified in subsequent printings if notice is given to the publisher.

The manufacturer's authorised representative in the EU for product safety is Oxford University Press España S.A. of el Parque Empresarial San Fernando de Henares, Avenida de Castilla,
2 – 28830 Madrid (www.oup.es/en).

# Contents

## 1  Geography ... and you   4
1.1  Welcome to geography!   4
1.2  What's in your geography kit?   5
1.3  How to get good at geography   6
1.4  Change in the Ironbridge Gorge   7
1.5  How to answer questions – part 1   8
1.6  How to answer questions – part 2   9
1.7  How to answer questions – part 3   10

## 2  Maps and mapping   11
2.1  Mapping through the ages   11
2.2  Plans and scale   12
2.3  The maps in your head   13
2.4  From an aerial photo to a map   14
2.5  Using grid references   15
2.6  How far?   16
2.7  Ordnance Survey maps   17
2.8  How high?   18
2.9  Where on Earth?   19

## 3  About the UK   20
3.1  Your island home   20
3.2  It's a jigsaw!   21
3.3  What's our weather like?   22
3.4  Who are we?   23
3.5  Where do we live?   24
3.6  How are we doing?   25
3.7  London, our capital city   26
3.8  Our links to the wider world   27

## 4  Glaciers   28
4.1  Your place ... 20 000 years ago!   28
4.2  Glaciers: what and where?   29
4.3  How do glaciers shape the land?   30
4.4  Landforms shaped by erosion – part 1   31
4.5  Landforms shaped by erosion – part 2   32
4.6  Landforms created by deposition   33
4.7  More about the Lake District   34
4.8  Do glaciers matter?   35

## 5  Rivers   36
5.1  Meet the River Thames   36
5.2  It's the water cycle at work   37
5.3  A closer look at a river   38
5.4  How do rivers shape the land?   39
5.5  Six landforms created by rivers   40
5.6  How do we use rivers?   41
5.7  What's the Thames Estuary like?   42
5.8  Floods!   43
5.9  Flooding on the River Thames   44
5.10 Can we protect ourselves from floods?   45

## 6  Africa   46
6.1  What and where is Africa?   46
6.2  A little history   47
6.3  What's Africa like today?   48
6.4  The countries in Africa   49
6.5  Population distribution in Africa   50
6.6  What are Africa's main physical features?   51
6.7  Africa's biomes   52

## 7  Kenya   53
7.1  Hello Kenya!   53
7.2  What are Kenya's main physical features?   54
7.3  What's Kenya's climate like?   55
7.4  A short history of Kenya   56
7.5  Kenya's population   57
7.6  What's Nairobi like?   58
7.7  What does everyone do?   59
7.8  How Kenya earns money from flowers   60
7.9  On safari!   61
7.10 So how is Kenya doing?   62

# Geography... and you

## 1.1 Welcome to geography!

**This is about introducing geography and the kind of things you will learn about.**

1  Geography can be divided into three different strands: *physical*, *human* and *environmental*. Explain what each one means in your own words.

   ................................................................................................................

   ................................................................................................................

   ................................................................................................................

2  **a**  Now write down any topics you can think of that are part of geography (try to think of at least eight).

   

   **b**  Circle any topics that are *physical* geography in one colour, *human* geography in another, and *environmental* geography in a third.

   **c**  Do any of your topics include all three strands? Circle them in a fourth colour.

3  **a**  Have a look at this photo. Brainstorm some questions you could ask about it. Can you think of six?

   ................................................................................................................

   ................................................................................................................

   ................................................................................................................

   ................................................................................................................

   ................................................................................................................

   ................................................................................................................

   **b**  Are you able to answer any of your questions? How would you find answers to the others?

   ................................................................................................................

   ................................................................................................................

# 1.2 What's in your geography kit?

**This is about the resources you will use to study geography.**

1. What are the advantages and disadvantages of paper maps and Google Maps? Complete the table below. Try and think of at least two for each box.

|  | Advantages | Disadvantages |
|---|---|---|
| Smart maps, e.g. Google Maps |  |  |
| Paper maps, e.g. OS maps |  |  |

2. Think about your local area. Imagine you had to take one photo to show someone living on the other side of the world what your area is like. Would you take a ground-level photo or an aerial photo? Why?

3. Draw a picture in the space below to show what your photo would look like.

**Tip!** Remember that your aim is to give as much information about your local area as possible.

Geography... and you

# 1.3 How to get good at geography

**This is about the skills you will need to become a good geographer.**

1. Choose one of the famous explorers on page 11 of *geog.1*. Create a fact file for them using the template below.

   Name: ......................................................................................

   Picture:

   Year of birth: ............................................................................

   Year of death: ...........................................................................

   How they were good at geography:

   ......................................................................................................................................................

   ......................................................................................................................................................

   ......................................................................................................................................................

   ......................................................................................................................................................

2. Now use the internet or your school library to research another famous explorer. Create a fact file for them.

   Name: ......................................................................................

   Picture:

   Year of birth: ............................................................................

   Year of death: ...........................................................................

   How they were good at geography:

   ......................................................................................................................................................

   ......................................................................................................................................................

   ......................................................................................................................................................

   ......................................................................................................................................................

# 1.4 Change in the Ironbridge Gorge

**Here you will look at an example of change in Shropshire.**

1 Look at these statements. They are all about the history of the Ironbridge Gorge, but they are jumbled up. Number them from 1–6 to put them in the correct order.

People began using the area's natural resources, such as coal, for fuel. ☐

The local people decided to build museums to celebrate their history and rejuvenate the area. ☐

Abraham Darby's method of extracting iron helped to kick-start the Industrial Revolution. ☐

People settled on the slopes of the gorge. ☐

The steep slopes meant it was hard to get to Ironbridge, and the area began to decline. ☐

A cast iron bridge was built across the river, and the town of Ironbridge grew nearby. ☐

2 What is the main industry in Ironbridge today? _____

3 Imagine you are in charge of attracting visitors to the Ironbridge Gorge. In the space below, design a poster to encourage people to visit the area.

**Tip!**
Think about the following:
- What is there to do?
- What can visitors see?
- What can they learn?

Geography... and you

# 1.5 How to answer questions – part 1

**Here you will look at the first set of command words in geography.**

1  **Complete** the sentence below.

   *Command words help you to…* _____

2  The following questions let you practise using command words.

   a  **Name** the county in which Ironbridge is found.

   _____

   b  The table below shows the populations of some places in the Ironbridge Gorge

   |  | Population (2011 Census) |
   |---|---|
   | Broseley | 4929 |
   | Ironbridge | 2582 |
   | Madeley | 5329 |

   i  **Name** the town with the smallest population.

   _____

   ii  **Calculate** the total number of people living in Ironbridge and Brosely combined.

   _____

   iii  **Calculate** how many more people live in Madeley than in Ironbridge.

   _____

   c  **Suggest** two reasons why people visit the Ironbridge Gorge museums.

   _____

   _____

3  Now think up three questions of your own using the command words in the box. They can be about your local area or another place you have learnt about in Geography.

   | calculate | copy and complete | define | give | identify | name | state | suggest |

   _____

   _____

   _____

8  Geography… and you

# 1.6 How to answer questions – part 2

**Here you will look at the second set of command words in geography.**

1 Look at the cartoons on page 16 of *geog.1*, showing how command words are used in everyday life. In the boxes below, draw your own examples for each command word.

| **Compare** | **Describe** |
|---|---|
|  |  |

| **Explain** |
|---|
|  |

| **Justify** | **Outline** |
|---|---|
|  |  |

Geography... and you

# 1.7 How to answer questions – part 3

**Here you will look at how to write a longer answer in Geography.**

1  Today, the Ironbridge Gorge receives over 500,000 visitors a year and generates millions of pounds for the economy. Use the framework below to help you answer the following question.

   *'The only reason tourists visit is to see the Iron Bridge.'*

   **To what extent** do you agree with this statement?

> **Tip!**
> Think about the following:
> - Why might tourists want to see the Iron Bridge?
> - What other reasons might tourists have to visit Ironbridge? (History, museums, beautiful scenery…)

Opening sentence:
..........................................................................................................
..........................................................................................................
..........................................................................................................

Arguments agreeing with the statement:
..........................................................................................................
..........................................................................................................
..........................................................................................................
..........................................................................................................
..........................................................................................................
..........................................................................................................

Arguments disagreeing with the statement:
..........................................................................................................
..........................................................................................................
..........................................................................................................
..........................................................................................................
..........................................................................................................
..........................................................................................................

Your opinion and conclusion:
..........................................................................................................
..........................................................................................................
..........................................................................................................
..........................................................................................................

Geography… and you

# Maps and mapping

## 2.1 Mapping through the ages

> Stick a map, or part of a map, in this box.
> It can be any sort of map – just as long as it's a map.
> Then answer the questions below.

**1** What does your map show?

.................................................................................................................................

**2** What's the best thing, or most interesting thing, about your map?

.................................................................................................................................

**3** How would you define a map? Finish this sentence:

A map is .........................................................................................................................

.................................................................................................................................

**4** Write down when, how, and why, you last used a map.

.................................................................................................................................

.................................................................................................................................

# 2.2 Plans and scale

**This is about what plans are, and what the scale of the plan tells you.**

1. A drawing of something seen from above is called a **plan**.

   Match the drawing of a chair with the correct plan (tick the correct one).

    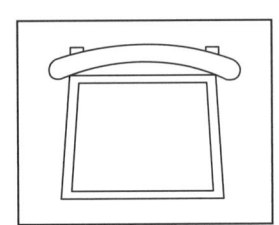 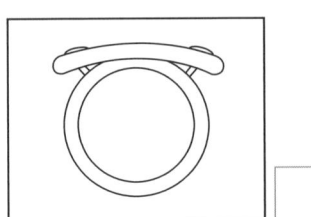 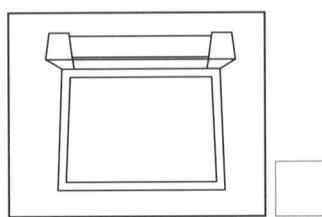

2. Now look at this plan of a bedroom. 1 cm on the plan represents 40 cm in the room. That is the **scale** of the plan. Use the plan to fill out the gaps in the sentences below.

   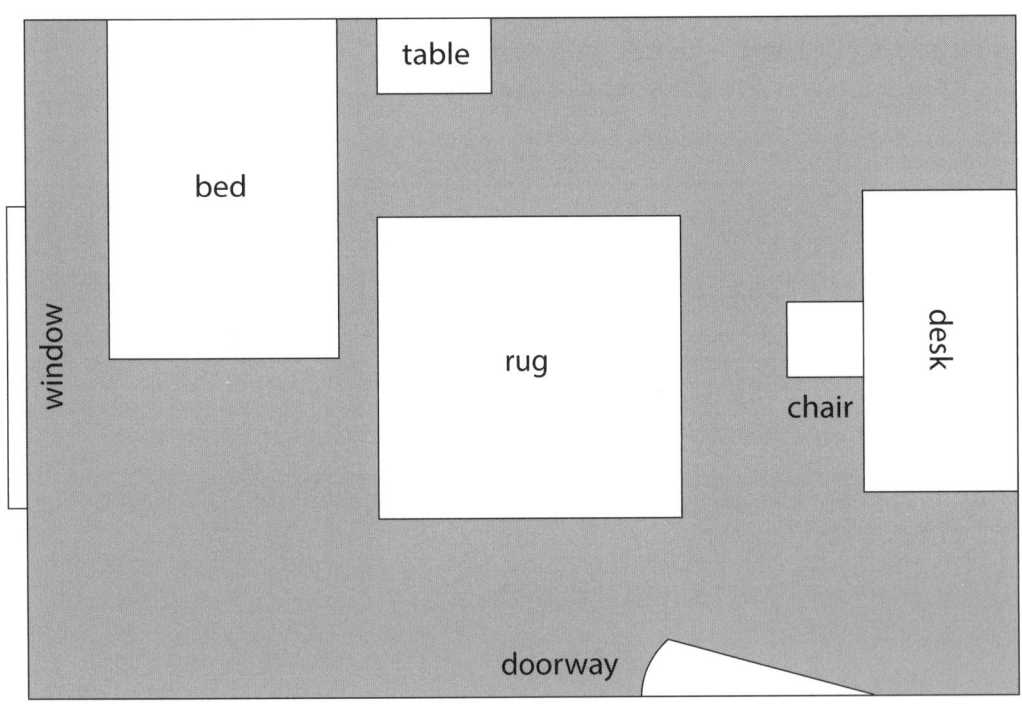
   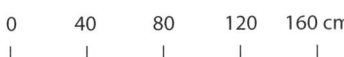

   a  On the plan, the window is this wide: ............

   So, in real life the window is ............ cm wide.

   b  Now measure the length of the bed and fill in the gaps below.

   On the plan, the bed is ............ cm long.

   This means it is ............ cm long in real life, which is ............ m.

   c  Something in the room is 60 cm wide in real life. What is it? ............

   d  What in the room is 160 cm × 80 cm in real life? ............

## 2.3 The maps in your head

**This is about your personal mental maps.**

1   **a**  Think about a place you know well – it may be your local area or a park, for example. In the space below, draw a mental map of that area.

**b** Add labels to show your feelings about the various parts of your map. Some of your feelings may be happiness, excitement, fear or sadness, but you may be able to think of others as well.

Think of a symbol for each of your feelings and draw the symbol in the correct place on your map.

**c** Show your map to a partner. Write down the thing that they like best about your map.

# 2.4 From an aerial photo to a map

**Here you will look at an aerial photo and draw a map of the same area.**

This photo shows a railway bridge over the River Tamar in Devon.
Your task is to draw a sketch map of the same place.
Don't forget a key!

**Key**

Maps and mapping

## 2.5 Using grid references

**This is about finding places on a map, using grid references.**

1  Fill in the gaps in this sentence, choosing from the words in the box.

   A good map has five things: a _____, a frame around it, an arrow to show _____, a _____ and a _____.

   | title | note |
   | north | scale |
   | east | lock |
   | key | river |

2  Look at this map.

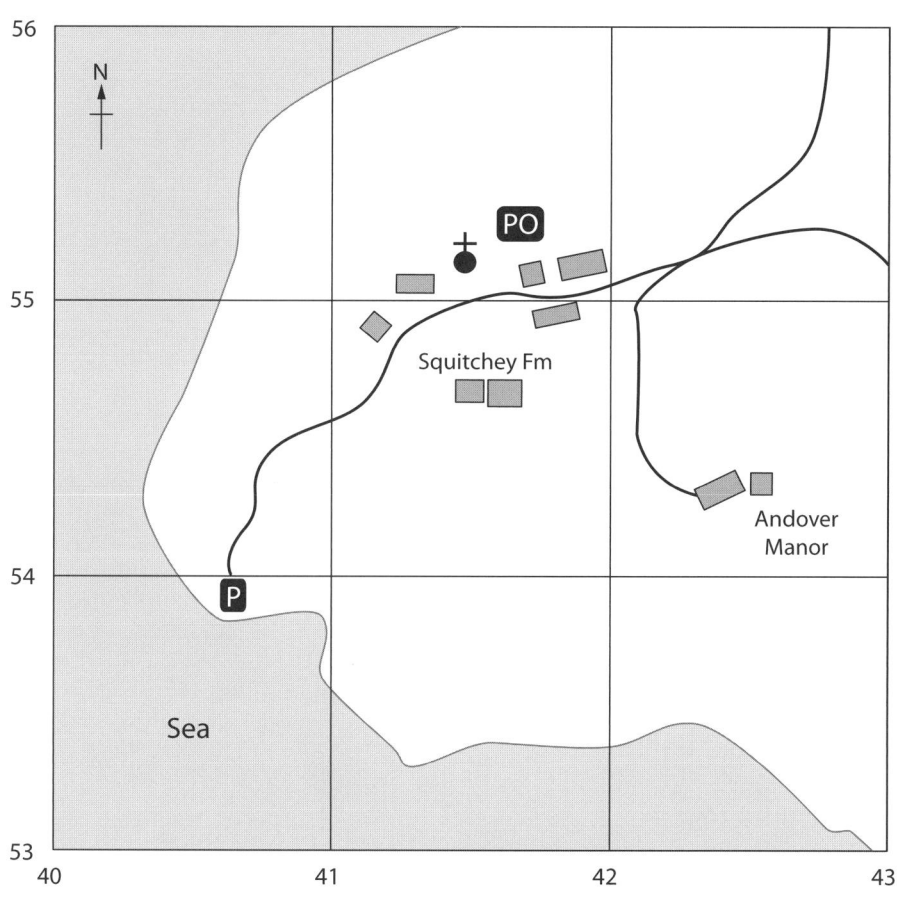

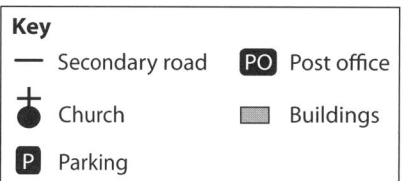

   Give a four-figure grid reference for:

   a  Squitchey Farm _____  b  Andover Manor _____  c  the church _____

3  What is at this grid reference on the map?

   a  407539 _____  b  414552 _____  c  416553 _____

4  Now add two more things to the map. Name them, and give their six-figure grid references.

   a  _____ is at _____

   b  _____ is at _____

Maps and mapping   15

# 2.6 How far?

**This is about how to find the distance between two places on a map.**

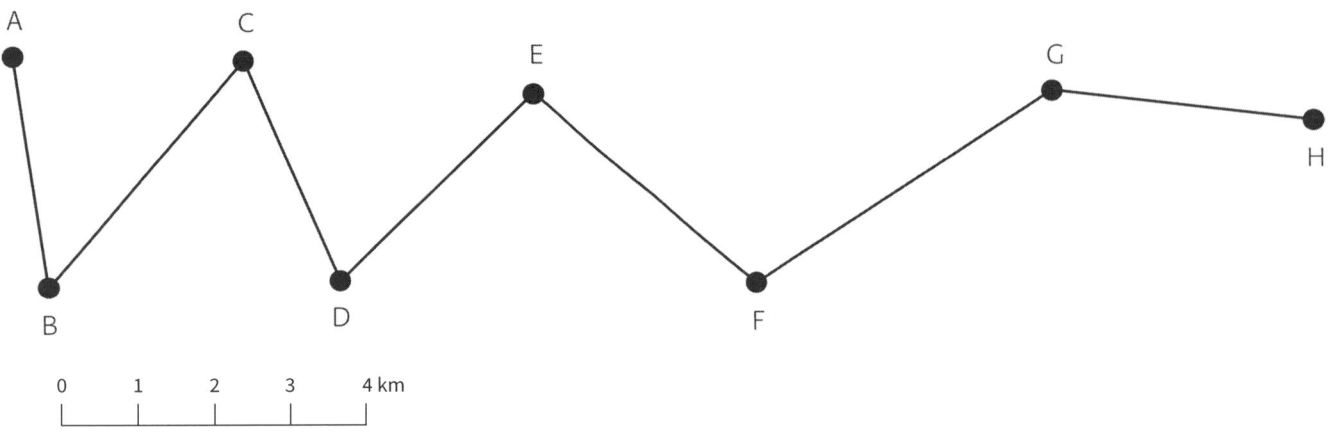

1. How far is it as the crow flies:

   a from A to C? ........................   b from A to H? ........................

2. How far is it by road:

   a from A to C? ........................   b from A to H? ........................

3. Have a look at the map on the page opposite.

   a How far is it as the crow flies from Hella Point (in square 3721) to Pordenack Point (square 3424)? ........................

   b How far is it by road between St Buryan (4125) and Trethewey (3823)?

   ........................

4. a Follow these instructions.

   Drive east from Land's End for just over a kilometre. Take a right turning and follow the road for 1.8 km. Turn left and follow the short track to the end.

   Where do you end up? ........................

   b Now give instructions (as in **a**) to someone who wants to travel from Treen (3923) to Trebehor (3724).

   ........................

   ........................

   ........................

## 2.7 Ordnance Survey maps

**This is about what OS maps are, what they show, and how to use them.**

This OS map shows part of the Land's End peninsula in Cornwall.

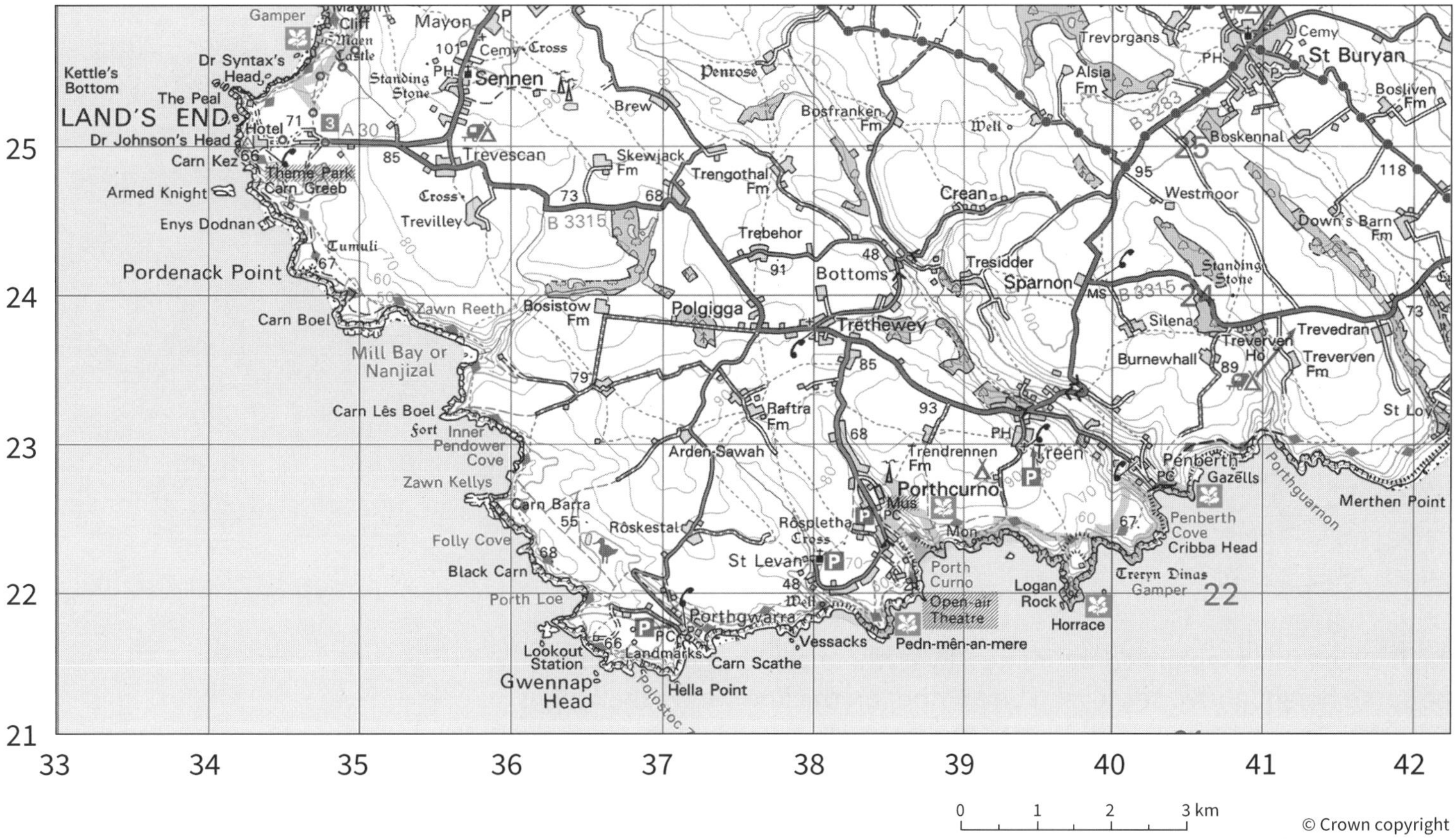

© Crown copyright

1  What is at each of these grid references?

   a  387219 ..................................................

   b  385253 (Hint: Fm means farm) ..................................................

   c  366249 ..................................................

   d  345242 ..................................................

2  Find one of each of these on the map and give a six-figure grid reference for it.

   a  a car park ..................................................

   b  a church ..................................................

   c  a public phone ..................................................

   d  a camp site ..................................................

3  What clues are there on the map that the Land's End peninsula gets lots of visitors?
   Give as many as you can.

Maps and mapping 17

# 2.8 How high?

**This is about how height is shown on an OS map.**

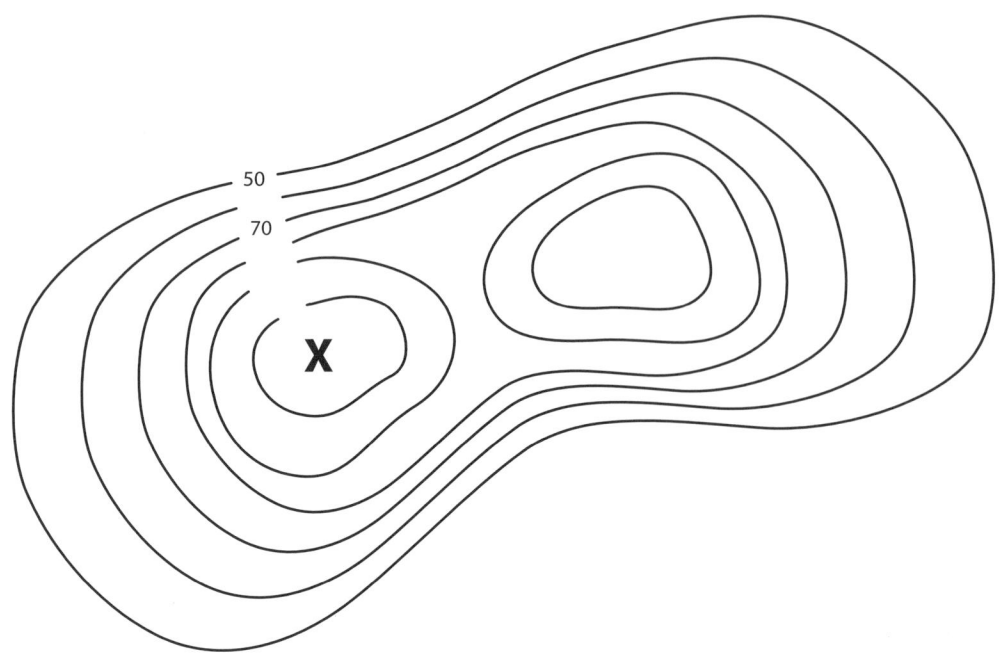

1  The lines on this map are contour lines. Everything along a contour line is the same height above sea level. The number on the line shows the height in metres. These contour lines are at 10 m intervals.

   a  Write in the four missing labels.

   b  Above what height is the land at X? ........................

   c  Colour in all the land above 80 metres.

   d  Write a label to show where the slope is steep.

   e  Now write a label to show where the slope is gentle.

2  Now look at the OS map on page 17 of this workbook. About how high above sea level is:

   a  Trevilley (358246)? ........................

   b  Raftra Farm (376233)? ........................

3  a  Complete this sentence:

      Another way that OS maps show how high a place is by using ........................
      ................................................. . These give the exact height at a specific point, in metres above sea level.

   b  Can you find an example of this on the OS map on page 17? Give a four-figure grid reference.

      ........................

18  Maps and mapping

# 2.9 Where on Earth?

**This is about the special grid lines we use to say where places are on Earth.**

1. Circle the correct word in each of these sentences.

   a. The lines that circle the Earth from top to bottom are lines of **longitude** / **latitude**.

   b. The lines that circle the Earth from side to side are lines of **longitude** / **latitude**.

   c. The 0° line of latitude is called the **Equator** / **Arctic Circle**.

   d. The 0° line of longitude is called the **Equator** / **Prime Meridian**.

2. Write these statements as coordinates:

   a. 33° north of the Equator, 20° east of the Prime Meridian. _____

   b. 44° south of the Equator, 40° west of the Prime Meridian. _____

3. Look at the map.

   a. Finish labelling the lines of latitude by filling in the gaps.

   b. Label the five main lines of latitude: the Equator, the Tropic of Cancer, the Tropic of Capricorn, the Arctic Circle, and the Antarctic Circle.

   c. Now finish labelling the lines of longitude.

   d. Label the Prime Meridian.

   e. Shade the area between the Tropic of Cancer and the Tropic of Capricorn in a 'warm' colour. Label this region 'the tropics'.

   f. Shade the Arctic and Antarctica in a 'cool' colour. Label these regions.

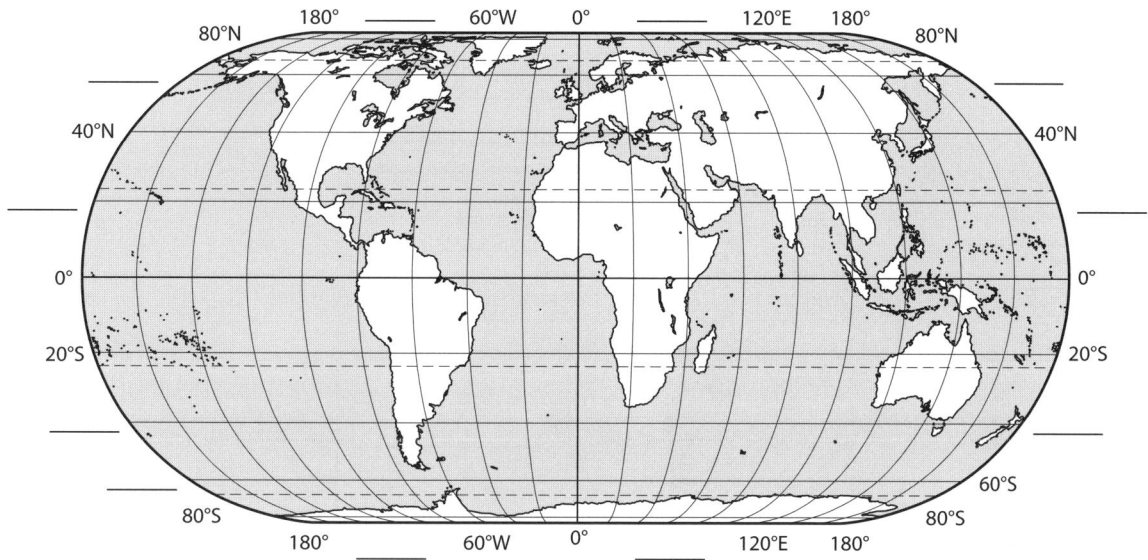

4. Mark three dots on the map, each one in a different continent. Label them X, Y, and Z. Write the coordinates of your places here:

   X _____ Y _____ Z _____

Maps and mapping   19

# About the UK

## 3.1 Your island home

**This is about the forces that shaped the British Isles – and about Britain's main physical features.**

1. This paragraph explains how the British Isles were formed. Choose words from the box to fill in the gaps.

   Once upon a time, the British Isles lay at the _____, as part of a giant _____. When this broke up, they drifted _____ as part of Europe. As they drifted, over millions of years, they went through many _____. They became desert. They were frozen in _____. They were drowned by the _____. They had earthquakes and eruptions. They got pushed and squeezed until _____ grew. And then they got _____ from the rest of Europe.

   | ice | cut off | changes | sea | crust |
   |---|---|---|---|---|
   | continent | mountains | currents | north | Equator |

2. Here are some features of the British Isles, but they're all jumbled up. Unscramble the words and then use them to label the map below.

   vrier sernve          verri nttre          rierv htmaes          alke drictist
   eisglhn nanchel       ninespen             rthno aes             rthno estw landshigh
   isirh sae

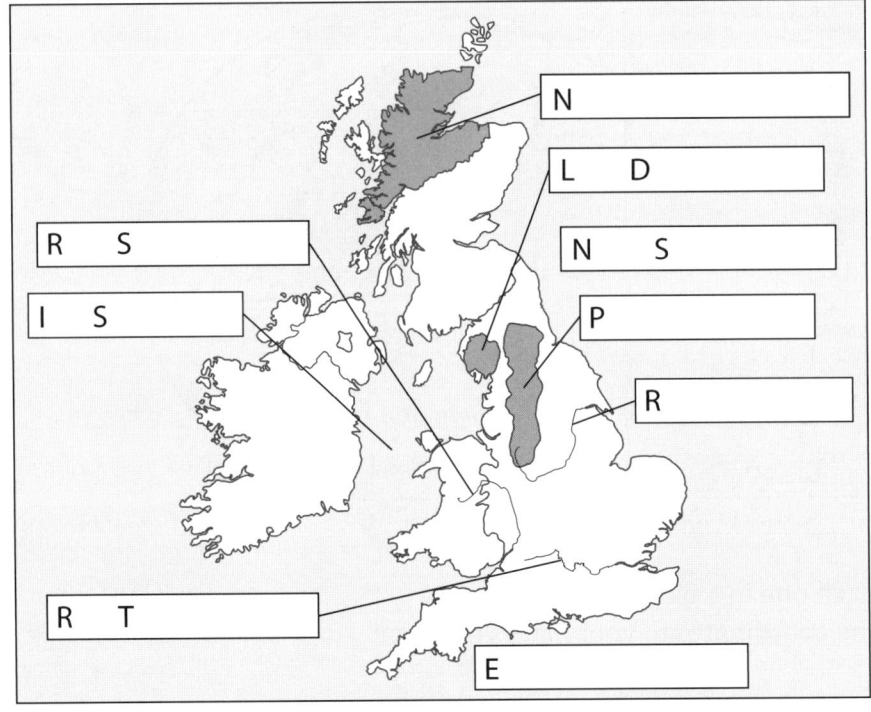

# 3.2 It's a jigsaw!

**This is about how we humans have carved up the British Isles.**

1. Fill in the gaps using words from the box.

   The British Isles is divided up into two countries: the United Kingdom and the

   _____ .

   The _____ in turn is made up of different nations: England,

   _____ , Wales and Northern Ireland.

   | United Kingdom | Germany | British Isles |
   |---|---|---|
   | Republic of Ireland | Scotland | England |

2. Now look at the map and answer these questions.

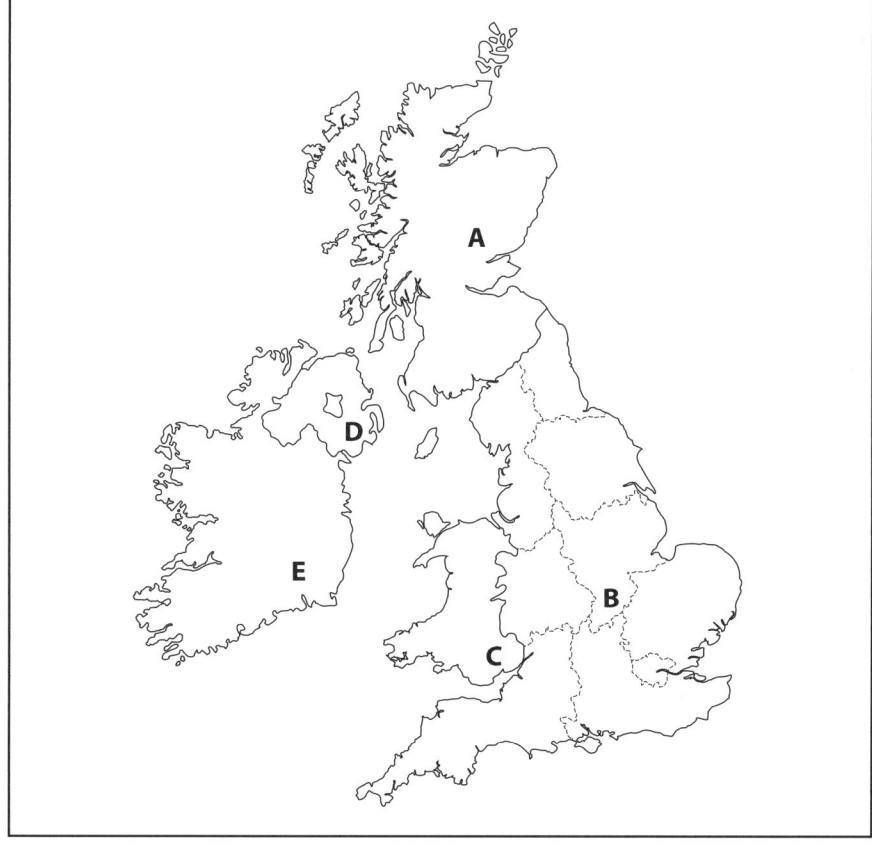

   a  A is called

   b  B is called

   c  D is called

   d  A–D together are called

   e  A–E together are called

   f  Finally, shade in the countries that make up Great Britain.

3. Draw a line on the map to the country or region that you live in and label it.
   What is special about your country or region?

About the UK  21

# 3.3 What's our weather like?

**This is about the difference between weather and climate – and how the climate varies across the UK.**

1  Fill in the gaps using the words in the box.

   a  ........................... means the state of the ........................... .
      Is it warm? wet? windy?

   b  ........................... is the ........................... weather in a place.

   | weather | average | atmosphere | climate |

2  a  Circle the correct word in these sentences.
      In general:

      i    It is **colder** / **warmer** in the north, because it is further from the Equator.

      ii   It is also **colder** / **warmer** on high land. Up a mountain the temperature **falls** / **rises**.

      iii  In winter, a **cold** / **warm** ocean current called the North Atlantic Drift **cools** / **warms** the west coast. So the east coast is the **coldest** / **warmest** part in winter.

   b  The maps below show average temperatures in summer and winter. Colour in the first map in shades of orange. Make the warmest areas darkest, and the coldest areas lightest.

   c  Now colour the second map in shades of blue. This time, make the coldest areas darkest. And don't forget the key!

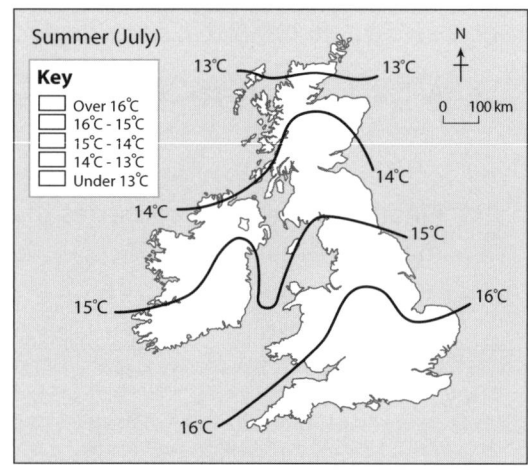
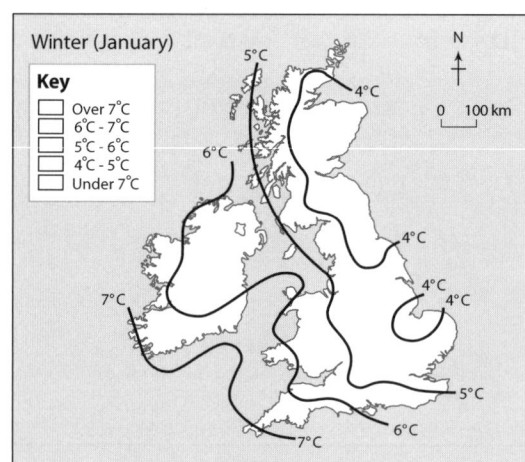

3  Answer these questions in full sentences.

   a  Which parts of the British Isles are wettest?

   ........................................................................

   b  Can you explain why?

   ........................................................................

# 3.4 Who are we?

**This is about how Britain has been peopled by immigrants.**

1. Draw lines to link the terms below with their definitions.

|  |  |
|---|---|
| asylum seeker | someone who takes over land to live on, where no-one has lived before |
| invader | someone who enters a country to attack it |
| refugee | someone who comes into a country to live |
| emigrant | someone who has been forced to flee from danger |
| settler | someone who moves to another part of the country or another country, often just to work for a while |
| immigrant | someone who leaves their own country to settle in another country |
| economic migrant | someone who flees to another country for safety, and asks to be allowed to stay there |

2. Here are some statements from people who've arrived in the British Isles in the last 2000 years.

> It's 48 CE. I am a centurion with the Roman army. Our aim is to expand our empire.

> I came to live here back in 2001. I didn't want to leave Kosovo but the war meant it was too dangerous to stay.

> It's 1956. I've come here from Jamaica in search of a job.

> It's 4000 BCE. I've come here from Europe with my tribe. We're looking for a good place to farm.

Choose what you think is the best term for each person (use terms from question 1).

3. 'In the British Isles, we are all immigrants.' Do you agree with this statement? Justify your answer.

# 3.5 Where do we live?

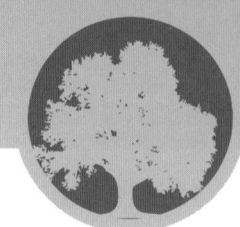

**This is about how we humans have shaped the country, through where we chose to live!**

1 Fill in the gaps.

The ............... ............... of a place is the average number of people per square kilometre.

2 Tick the correct answer.

   a The nation with the highest population density is …

   England ☐    Wales ☐    Scotland ☐

   b Of these areas, the one with the lowest population density is …

   Cumbria ☐    Greater Manchester ☐    Devon ☐

   c Of these cities, the largest is …

   Edinburgh ☐    Birmingham ☐    Glasgow ☐

3 Look at this pie chart for the United Kingdom.

   **Where the UK population lives**

   Key
   ☐ urban areas
   ☐ rural areas

   a Shade in the chart and the key to show where the population of the UK lives.

   b Imagine you live on a farm half an hour's drive to the nearest town.
   Give three good points and three bad points about living in such a rural area.

   **Good points**

   ......................................................................................
   ......................................................................................
   ......................................................................................

   **Bad points**

   ......................................................................................
   ......................................................................................
   ......................................................................................

# 3.6 How are we doing?

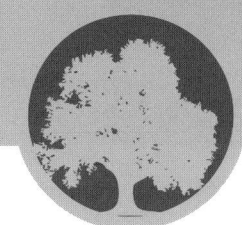

**This is about how some parts of Great Britain are better off than others – and some of the reasons why.**

House prices, wages and unemployment vary from region to region in the UK.

| Region | Average house price (£) | Average wage (£ per week) | Unemployment (%) |
|---|---|---|---|
| A Wales | 163 902 | 509 | 3.8 |
| B Scotland | 150 825 | 563 | 3.3 |
| C Northern Ireland | 134 811 | 521 | 3.1 |
| D North East | 130 888 | 507 | 5.6 |
| E North West | 161 891 | 530 | 4.1 |
| F Yorks & Humber | 161 443 | 521 | 5.0 |
| G East Midlands | 192 682 | 516 | 4.2 |
| H West Midlands | 195 498 | 537 | 4.8 |
| I East | 289 476 | 558 | 2.8 |
| J London | 471 504 | 713 | 4.3 |
| K South East | 318 727 | 589 | 2.8 |
| L South West | 253 410 | 531 | 2.6 |

* 2018–9 figures

1  Complete the bar chart below for house prices and wages in 2018–9. Wales has been done for you.

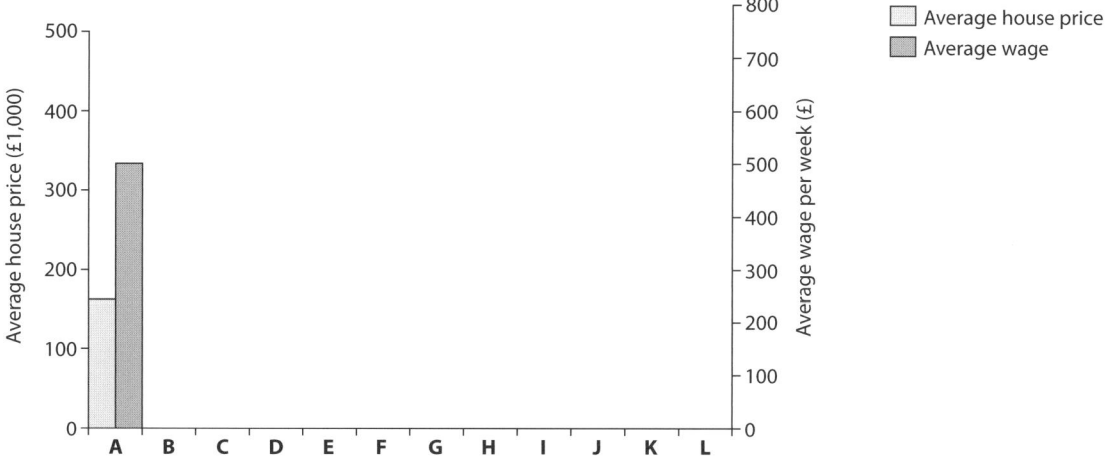

2  What do the differences in house prices tell you about inequality in the UK?

...................................................................................................................................................

...................................................................................................................................................

3  What does the relationship between house prices and wages tell you about inequality in the UK?

...................................................................................................................................................

...................................................................................................................................................

4  With the unemployment figures in mind as well, which region of the UK do you think would be the best to live in for income, employment and affordable housing? Why?

# 3.7 London, our capital city

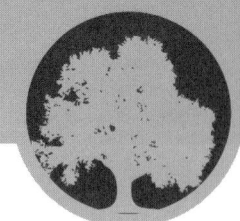

**This is about the London marathon. Can you work out its route from a description?**

The London marathon is the most popular marathon in the world. In 2019, over 40,000 people took part.

1. Read this description of the route of one year's London Marathon. Then trace the route of the race on the map below. Write in the words 'Start' and 'Finish'. Write in the missing words in the description and insert the numbers in the correct places on the map.

The race starts in Greenwich Park, south of the river and just to the east of the well-known loop of the Thames at the Isle of Dogs. The runners head east, across the A2 to _____ Park. (1) At the A _____ (2) road the race turns north, towards the river and then turns west along _____ Church Street (3). They pass Maryon _____ (4) and run along the A206, over the A2 again and towards Greenwich once more where they run past the famous ship, the Cutty _____ (5). They then turn north-west, until they get to the old docks at Surrey _____ (6). They then follow the loop of the river past the _____ Tunnel (7). Heading west along the river past Bermondsey, they cross the Thames at _____ Bridge (8). They now head east again past Wapping, and turn south into the Isle of Dogs, running along the Westferry Road. At the southernmost point of the loop they turn north, running towards the well-known skyscraper of Canary _____ (9) Near the _____ Tunnel (10) they turn west, heading for Limehouse. They then follow the Thames. They are now in the old City of _____ (11). Still heading west they run past Waterloo Bridge. From here they can see the National Theatre and the Royal Festival Hall on the _____ (12). At _____ near the Houses of Parliament, they head away from the river towards St James's Park. They run clockwise around the lake and the race ends at the northern end of the park.

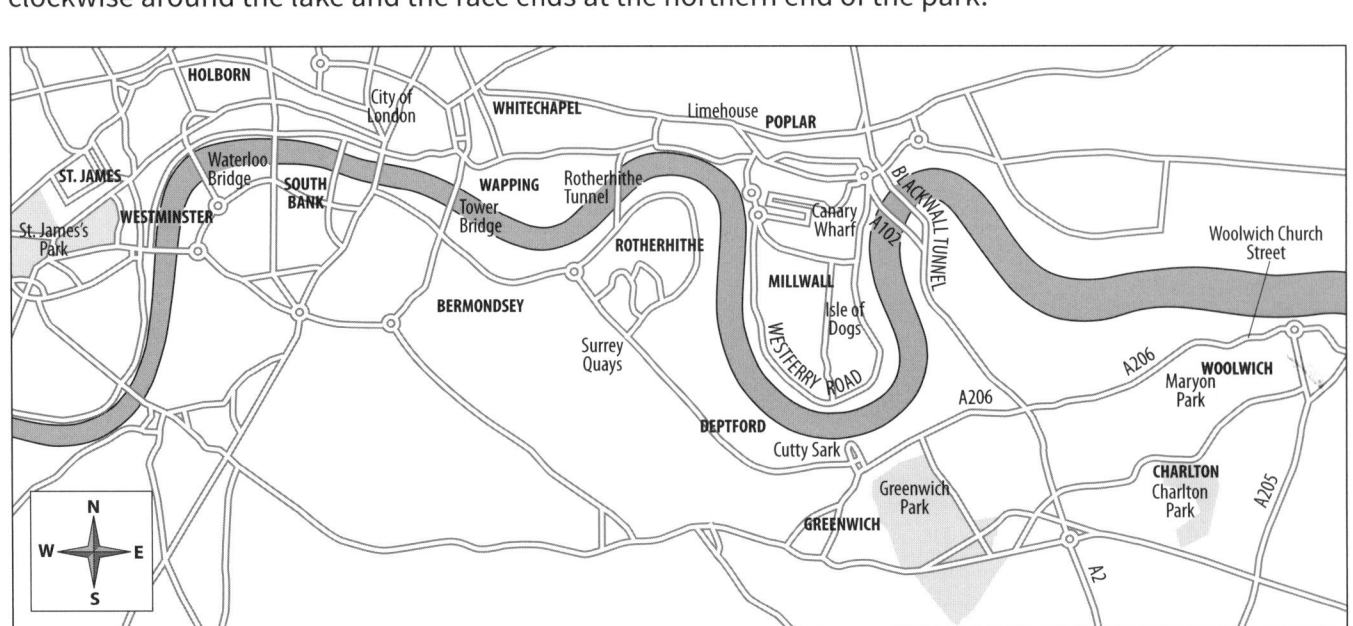

26  About the UK

# 3.8 Our links to the wider world

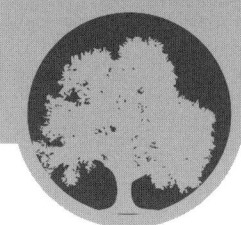

**Here you will examine how the UK is linked to the rest of the world.**

1  Match these links with examples of how they benefit the UK.

| Links | Examples |
|---|---|
| Trade | UK goods sold overseas make a profit and earn money |
| Transport | People in the UK can call or message friends and family all over the world |
| Communications | Tourists visit the UK and spend money in shops, hotels and restaurants |
| Tourism | People around the world listen to British music and watch British films |
| Culture | This link allows goods to be shipped to and from the UK, such as imported foods |

2

*Aid doesn't benefit the UK, so we should just stop paying it.*

**Tip!**
Think about the following:
- How aid helps other countries to develop
- The UK's global status
- What aid money is used for

Do you agree with this statement? Justify your answer.

................................................................................................................
................................................................................................................
................................................................................................................
................................................................................................................
................................................................................................................

About the UK

# Glaciers

## 4.1 Your place ... 20 000 years ago!

**This is about understanding when the British Isles was in the grip of ice.**

1 Write 'True' or 'False' in the box after each of these sentences.

   a  Woolly mammoths once roamed southern Britain.

   b  Woolly mammoths were like very large sheep.

   c  A few woolly mammoths can still be found in remote parts of Scotland.

   d  Woolly mammoths were like hairy elephants.

2 Finish this timeline. The 'time' labels are already in place. To finish it, you need to add notes at each time label, saying what was happening at that time.

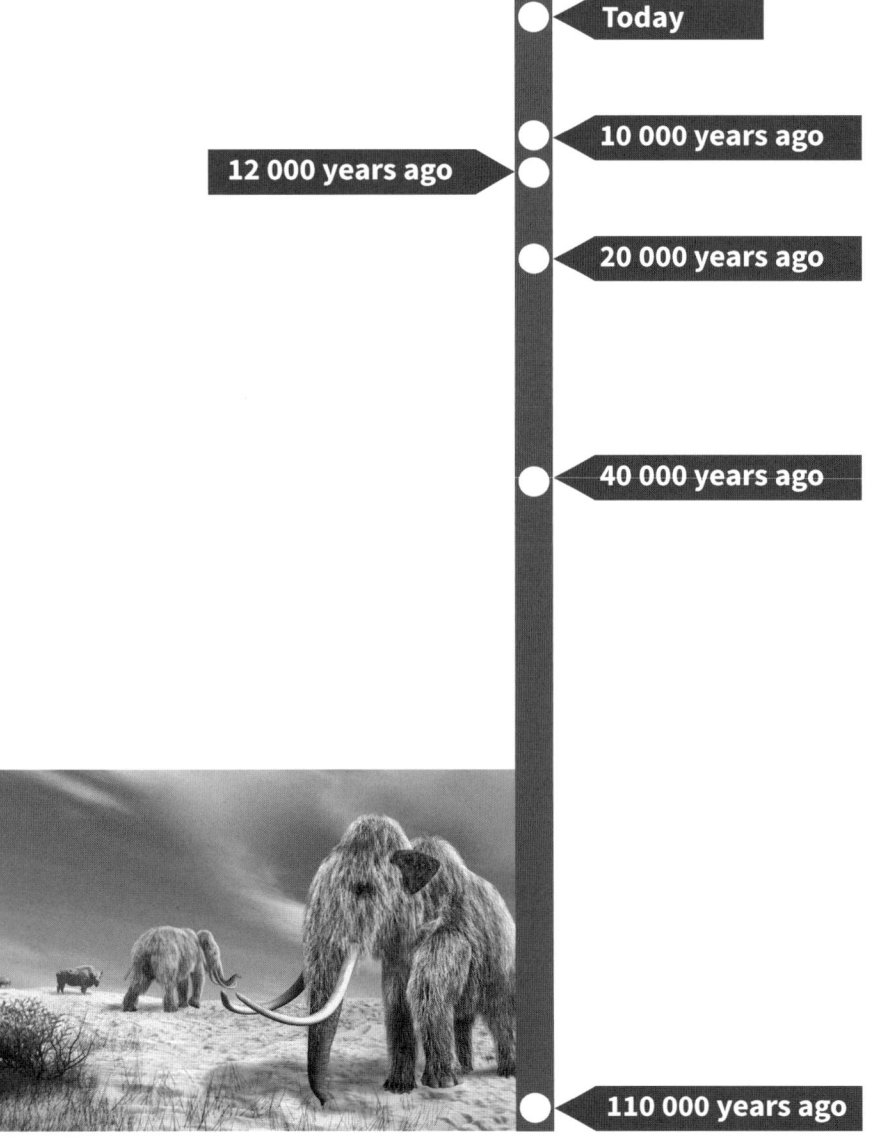

**Tip!**
Plan what you're going to write, before you start writing – and then write neatly! You could draw a box round your notes at each time, to help make your timeline extra clear.

28  Glaciers

# 4.2 Glaciers: what and where?

**This is about the world's glaciers today.**

1. Tick the correct answers to these questions:

   a How much of the Earth's surface do glaciers cover?

   about 40% ☐    about 30% ☐    about 20% ☐    about 10% ☐

   b During the last ice age, how much of the Earth's surface was covered by glaciers?

   about 43% ☐    about 33% ☐    about 25% ☐    about 20% ☐

   c Today, how much of the world's ice is in Antarctica and Greenland?

   less than 90% ☐    95% ☐    99% ☐    over 99% ☐

   d What are large cracks in glaciers are called?

   cravats ☐    crevasses ☐    crevices ☐    creases ☐

   e How many continents have glaciers?

   two ☐    three ☐    five ☐    all seven ☐

2. Do some research to find out about Vatnajökull Glacier in Iceland.

   Find a photo of Vatnajökull Glacier and stick it in the big box below, and then write a fact about Vatnajökull in each of the smaller boxes.

Glaciers 29

# 4.3 How do glaciers shape the land?

**This is your chance to show that you know how glaciers shape the landscape.**

1. Draw a spider diagram to show the work that glaciers do and how they do it. The first one has been started for you.

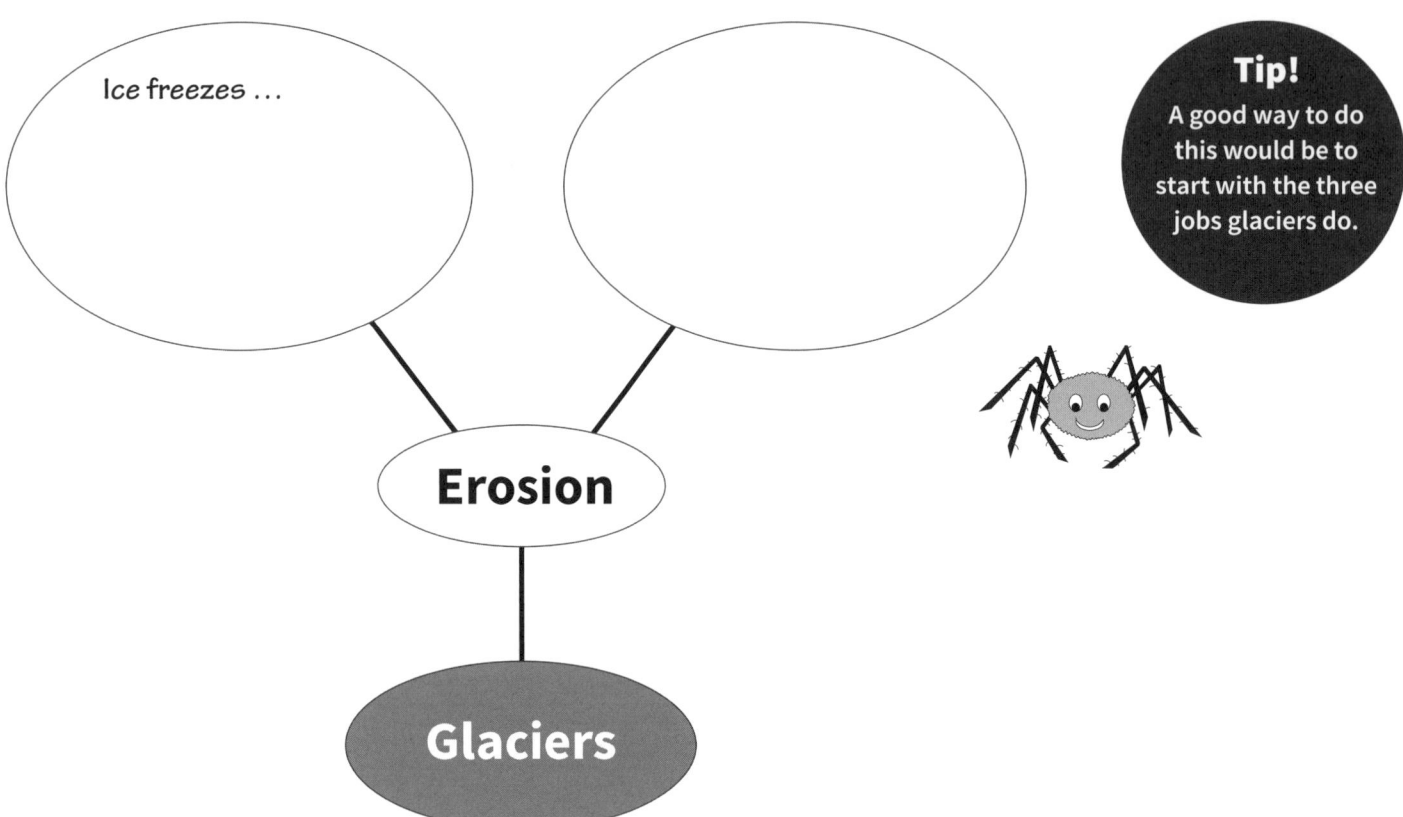

**Tip!** A good way to do this would be to start with the three jobs glaciers do.

**Tip!** You could colour-code the different parts of your spider diagram – this would help make it even clearer.

30  Glaciers

# 4.4 Landforms shaped by erosion – part 1

**This is your chance to learn how glaciers can change the landscape.**

1. The paragraph below describes how a corrie is formed.
   Circle the correct word from each pair.

   As snow falls, it **compacts** / **constructs** into ice. Through **abrasion** / **corrosion**, the hollow gradually becomes **bigger** / **smaller** and the walls steeper. Eventually, with the help of freeze-thaw weathering, the glacier is big enough to flow over the **edge** / **bottom** of the corrie and move down the mountain. When the glacier melts, the corrie may have a **lake** / **river** in it. This is often called a tarn.

2. Imagine you are going on an expedition to explore Bleaberry Tarn in the Lake District, shown in photo B on page 66 in *geog.1*. Write down five things that you think you would need to take to help make your expedition a success.

   Give a reason for each of your answers.

   | I would need to take… | because… |
   |---|---|
   | 1 | |
   | 2 | |
   | 3 | |
   | 4 | |
   | 5 | |

3. Study the photograph of Bleaberry Tarn again. Imagine you work for the Lake District Tourism Board. You have been asked to write about fifty words to describe what the photograph shows.

**Tip!** Remember to use as many adjectives (describing words) as you can.

Glaciers

# 4.5 Landforms shaped by erosion – part 2

**Here you will look at two more landforms created by glaciers.**

1. Look at photos A and B. Write down three differences between the valleys shown in each photo.

**Tip!** Think about the shape of the valleys.

**A**

**B**

1 ................................................................................................

2 ................................................................................................

3 ................................................................................................

2. Look at the map of the Lake District on the right. In your own words, give three facts about the lakes and where they are found. Remember to use the north arrow to help you.

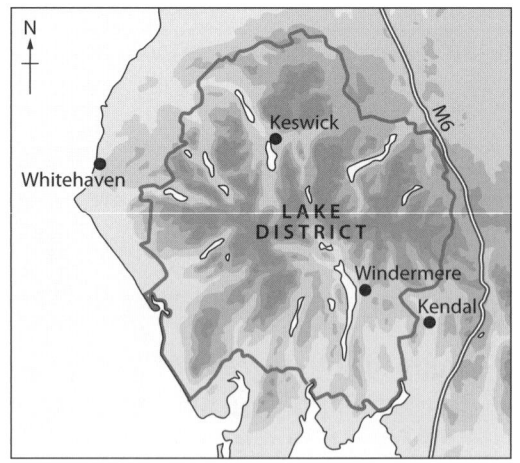

1 ................................................................................................

2 ................................................................................................

3 ................................................................................................

Glaciers

## 4.6 Landforms created by deposition

**This is where you learn about landforms created when a glacier melts.**

1. Glaciers form many landforms, even when they melt! Complete the spider diagram below that describes and explains the landforms that are created.

- A terminal moraine is…
  
  It is caused when…

- An erratic is…
  
  They are found in strange places because…

- **When glaciers melt they change the landscape**

- Drumlins are…
  
  They are smooth-shaped because…

- A lateral moraine is…
  
  It is caused when…

2. Write and present a two minute talk for your class about one of the landforms created by glaciers – either through glacial erosion or deposition. You should produce one diagram or PowerPoint slide to help you.

Glaciers

# 4.7 More about the Lake District

**Here you will look at the Lake District on an OS map.**

1  Here is a description of a place shown on the map in *geog.1* on page 73. Where is it? Use the clues to help you.

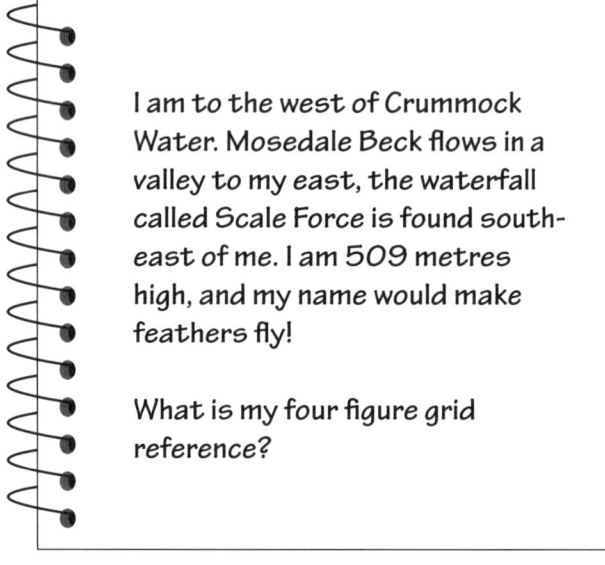

I am to the west of Crummock Water. Mosedale Beck flows in a valley to my east, the waterfall called Scale Force is found south-east of me. I am 509 metres high, and my name would make feathers fly!

What is my four figure grid reference?

The place is called _____

The four figure grid reference for it is _____

2  Now write your own clues to describe a place or feature shown on the map. Then test them on a partner. Can they identify it using your clues?

3  Use the clues below to follow a route around the area shown by the map. Write the places that you reach in the blank spaces in the paragraph. Remember to use the clues and the scale to help you!

I start walking from the car park at Gatesgarth at 6-figure grid reference

_____. I follow the road north-west until I get to the hotel in the

village of _____ in grid square 1716. Here, I take the footpath

on my left that crosses the valley floor and goes through the forest called

_____ Wood. I go past Bleaberry Tarn and climb upwards towards

_____ Pike. Here, I turn right and follow the long footpath north-west

until I get to a waterfall, called _____ at 6-figure

grid reference _____.

Glaciers

# 4.8 Do glaciers matter?

**This is your chance to consider whether glaciers really matter.**

1 Pages 74 and 75 in *geog.1* tell us about five facts about glaciers. Look at the list below and put them in rank order of importance, with 1 being the most important.

| Facts about glaciers | Rank order |
| --- | --- |
| Glaciers bring in tourists. | |
| Glaciers present a challenge. | |
| Glaciers support life. | |
| Glaciers are in need of protection. | |
| Glaciers warn us about climate change. | |

2 Write three sentences explaining why you have chosen your number 1. Remember to give reasons for choosing it!

Sentence 1

Sentence 2

Sentence 3

3 Design a poster in the space below showing what you know about glaciers. Use all the information that you have learned in this unit to help you.

Glaciers

# Rivers

## 5.1 Meet the River Thames

**This is where you will learn more about changes along England's longest river.**

1. The River Thames starts its life at Thames Head in the Cotswolds. Use the information on page 78 of *geog.1* and your own research to describe what the river is like as it flows through each of the other places shown on the map below. Complete the table below.

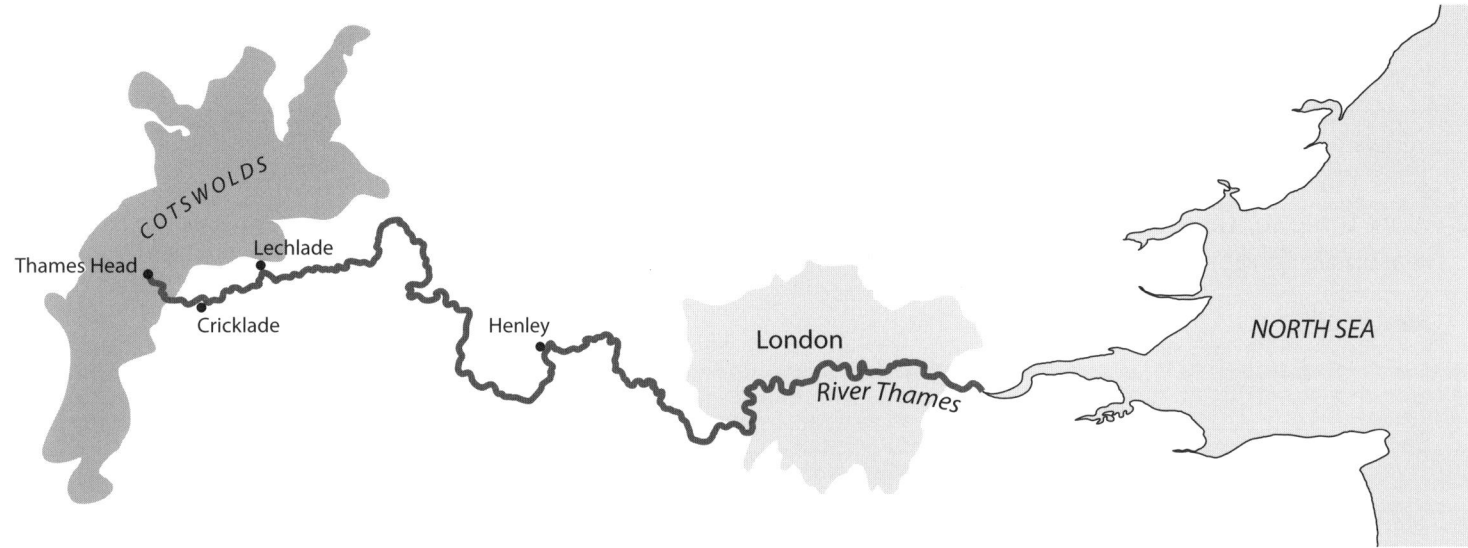

| When the River Thames flows through … | … it is like this: |
|---|---|
| Cricklade | |
| Lechlade | |
| Henley | |

# 5.2 It's the water cycle at work

**This is about the water cycle, and why it is important.**

1 Water moves between the ocean, the air and the land. This is called the water cycle.

   a Fill in the gaps below, choosing words from the box (you don't have to use them all).

   ☐ The air ............... . High up, where it's cooler, the water vapour ............... into tiny water droplets. These form ............... .

   ☐ The water drops fall as rain (or hail or sleet or snow).

   ☐ The sun warms oceans, lakes and seas, turning water into water vapour. This is called ............... .

   ☐ Some water runs along the ground, and some soaks through it, heading for streams and rivers.

   ☐ The droplets inside the clouds grow into larger droplets, leading to ............... .

   ☐ The river carries the water back to the ............... . The ............... is complete. And then it starts all over again…

   | evaporation | rises | gas | precipitation |
   | clouds | condenses | ocean | cycle |

   b Now write the numbers 1–6 in the small boxes to put the sentences in the correct order.

2 Draw a diagram in the box below to show how rainwater reaches a river. Try to use as many of these words as possible (there are a few clues to help you!):

interception (when leaves catch rainwater)
surface runoff
throughflow
groundwater

groundwater flow
infiltration (when rainwater soaks into the ground)
permeable (lets water soak through)
impermeable

Rivers 37

## 5.3 A closer look at a river

**This is about the different parts of a river.**

1  a  Circle the correct word in these sentences.

   i    The point where two rivers join is called a **tributary** / **confluence**.

   ii   The **confluence** / **watershed** is an imaginary line that separates one drainage basin from the next.

   iii  The **source** / **mouth** is where the river flows into a lake, or the sea, or the ocean.

   iv   The flat land around a river that gets flooded when the river overflows is the **floodplain** / **tributary**.

   v    The **mouth** / **source** is the starting point of the river.

   vi   The land around a river from which water drains into the river is the **river basin** / **watershed**.

   b  Now draw a sketch map of an imaginary river. Try to mark on and label all the features from **1a**.

2  Fill in the gaps to complete this diagram.

   **A drawing of the river's** _____

   - The _____ is the highest point.
   - The slope gets less steep in this middle stretch.
   - The _____ is the river's lowest point.
   - different layers of rock below the river
   - Now the slope is flattening out.
   - lake or sea

38  Rivers

# 5.4 How do rivers shape the land?

**This is about how rivers shape the land by picking up, carrying and dropping material.**

1. Fill in the gaps using words from the box.

   Rivers do their work in three stages:

   **1** They pick up or ................ material from one place.
   
   **2** They carry or ................ it to another place
   
   **3** Then they drop or ................ it.

   | erode | transport | deposit |
   | --- | --- | --- |

2. Finish off this cartoon to tell the story of Sid the Stone's journey. (You don't have to fill all the boxes.)

   **1** Sid the stone had lived in the river bank for as long as he could remember. Then, one day …

   **2** … he was prised out of the bank by **hydraulic action!**

   **3**

   **4**

   **5**

   **6**

Rivers 39

# 5.5 Six landforms created by rivers

This is about the landforms created by rivers as they erode and deposit material.

1 Fill in the gaps in this table.

| Landform | Description | Created by ... |
|---|---|---|
| V-shaped valley | a valley shaped like the letter V, carved out by a river | |
| waterfall | | erosion |
| gorge | a narrow valley with steep sides | erosion |
| | a bend in a river | erosion + deposition |
| oxbow lake | a lake formed when a loop of river gets cut away | |
| | ridges of hard rock in a valley that are eroded more slowly and stick out | erosion |

2 a These pictures show how a waterfall develops. Under each picture, describe what is going on.

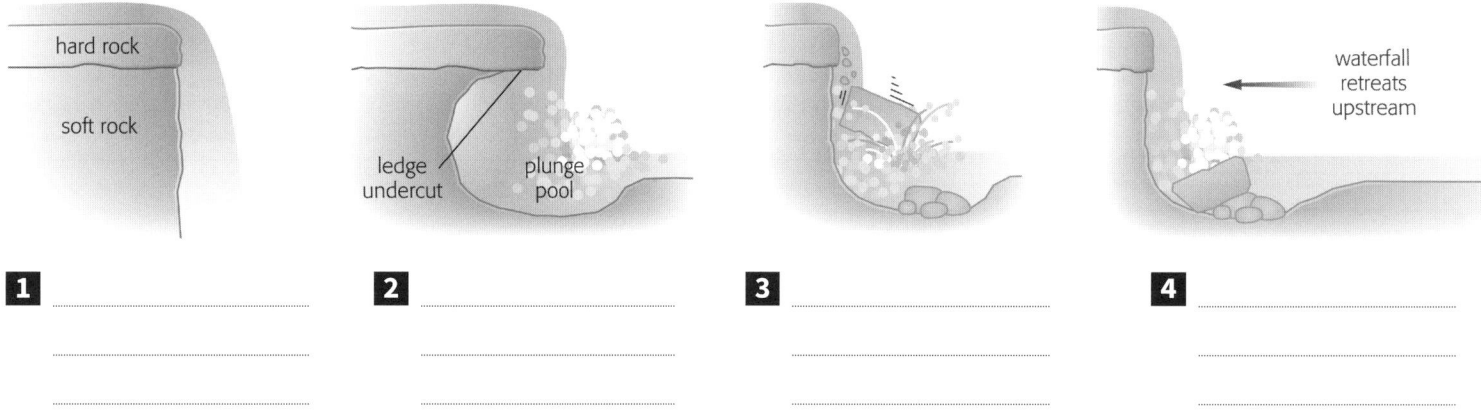

**1** ..................
**2** ..................
**3** ..................
**4** ..................

b Draw pictures in the boxes below to show how a meander develops.

**1** Water flows faster on the outer curve of the bend, and slower on the inner curve. So ...

**2** ... the outer bank gets eroded, but material is deposited at the inner bank. Over time ...

**3** ... as the outer bank wears away, and the inner one grows, a meander forms.

**4** As the process continues, the meander grows more 'loopy'.

40 Rivers

# 5.6 How do we use rivers?

**This is about how we make use of rivers and how we damage them.**

1  a  Look at the statements below. Draw lines linking the statements that you think are connected. One has been done for you.

   A  Dams are built across some rivers          and then pumped to our taps.

   B  Farmers pump water from rivers             and then sent back to rivers.

   C  Water from rivers is cleaned               to irrigate their land.

   D  Dirty water from our houses is cleaned     so that the water turns turbines to make electricity.

   b  Choose one of your completed statements from A to D. Copy it out on the line below.

   ....................................................................................................

   c  Do you think what you have just written is good or bad for the river? Give reasons for your answer.

   ....................................................................................................
   ....................................................................................................
   ....................................................................................................

2  Read the speech bubbles below.

*Rivers are for wildlife more than for people.*
Environmentalist

*My fishing lines get caught by the boats and break.*
Fisherman

*Without the water from the river my crops will not have enough water.*
Farmer

*We are quiet and do not disturb anybody.*
Boat owner

*There are very few places left where we can walk our dogs in peace*
Dog walker

Choose the one person that you think may damage the river the most. Give reasons to explain your choice.

....................................................................................................
....................................................................................................

## 5.7 What's the Thames Estuary like?

**Here you will look at the Thames Estuary and why it is important.**

1 Complete these photo captions to explain how the Thames Estuary is important.

| This photo shows… | This photo shows… | This photo shows… |
| --- | --- | --- |
| It is important because… | It is important because… | It is important because… |

2 Sara the Seal lives in the Thames Estuary. One day, she is swimming from Canvey Island towards the North Sea. Write a description of her journey. What does she see? What can she hear?

**Tip!** Use the photos and information on pages 90–91 of *geog.1* to help you.

Rivers

# 5.8 Floods!

**This is where you will find out how and why rivers flood.**

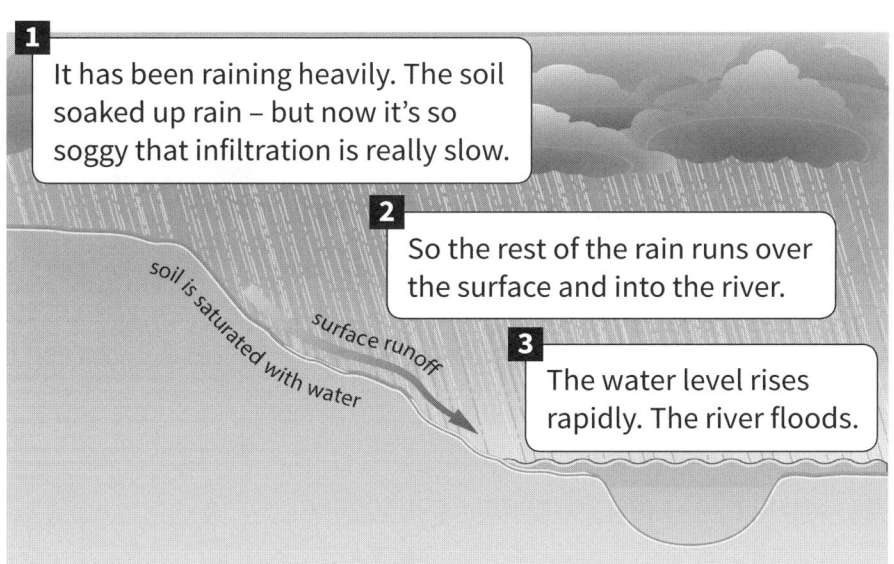

1. It has been raining heavily. The soil soaked up rain – but now it's so soggy that infiltration is really slow.
2. So the rest of the rain runs over the surface and into the river.
3. The water level rises rapidly. The river floods.

**1 a** Look at the diagram above. In your own words explain what is meant by the following words and phrases.

Infiltration: ............................................................................................................................

............................................................................................................................

............................................................................................................................

Surface runoff: ............................................................................................................................

............................................................................................................................

............................................................................................................................

**2** When floods happen people often talk about three things: flood prevention, flood defence and flood warning. In the spaces below write down what you think is meant by each term.

Flood prevention: ............................................................................................................................

Flood defence: ............................................................................................................................

Flood warning: ............................................................................................................................

**3** Which of the three do you think is the most important? Give reasons why.

I think .................................... is the most important because ....................................

............................................................................................................................

............................................................................................................................

Rivers

## 5.9 Flooding on the River Thames

**Here you will think about the impacts of flooding on the River Thames.**

1. Look at the photographs on page 94 in *geog.1* that show flooding in Oxford, Datchet and Richmond. Imagine that was your house… and your local area!

   In the space below, describe how you and your family would cope with the floods.

   **Tip!** Think about how it would change your daily life.

2. Write down five things that you think it would be essential to save if your house got flooded. Explain your answers.

   | I would save… | because… |
   |---|---|
   | 1 | |
   | 2 | |
   | 3 | |
   | 4 | |
   | 5 | |

3. **a** Look at the statements below that show what people living by the River Thames might do if their area was flooded. Put them in priority order, by writing the numbers 1–5 by each. Number 1 would be the first thing that you think they should do, and 5 the last.

   | Ring family | | Phone 999 | | Turn off power | | Save pets | | Move upstairs | |

   **b** In the space below, give reasons to explain your first choice.

   My first choice is

   This is because

44 Rivers

# 5.10 Can we protect ourselves from floods?

**This is where you will think about ways to reduce flood risk and protect ourselves from floods.**

1. Read this information about London. Do you think London should be protected by another Thames Barrier? Fill in the speech bubbles below to show your thoughts.

- London's population is 8.9 million (2018 data) and growing.
- In 2007, architect Sir Terry Farrell proposed building a five–mile stretch of islands along the mouth of the Thames, which would enhance the area's natural beauty and divert flood waters towards less built-up areas. The islands would be made up of London's waste and would use this to generate electricity.
- A second barrier could house turbines and use tide flows to generate electricity for London.
- Another idea is to set aside large areas of open country downstream of London as emergency floodplains, so protecting the capital.
- London's exports add up to about 27% of the total value of all UK exports.
- The Thames Barrier took eight years to build, costing £535m (£1.9b at 2019 prices) and became fully working in 1982.
- Europe's largest shopping centre is in London - Westfield in Stratford, near the Olympic Park.
- "The Thames Barrier was built in response to the floods in 1953. Nobody had heard of global warming then." Dr Richard Bloore (January 2013)
- In 2007, a second Thames barrier was costed at £20 billion.
- The Thames Barrier was originally designed to work until the year 2030.
- There are 270 stations on London's underground tube network.
- Over 20 million tourists visit London each year.
- The average house price in London (April 2019) is £471,504.
- Climate change is causing sea level rise, which could mean larger storm surges in the Thames.
- The Thames Barrier currently protects 125sq km of London, including an estimated 1.25m people, £80bn worth of property, a large proportion of the London tube network and many historic buildings, power supplies, hospitals and schools.

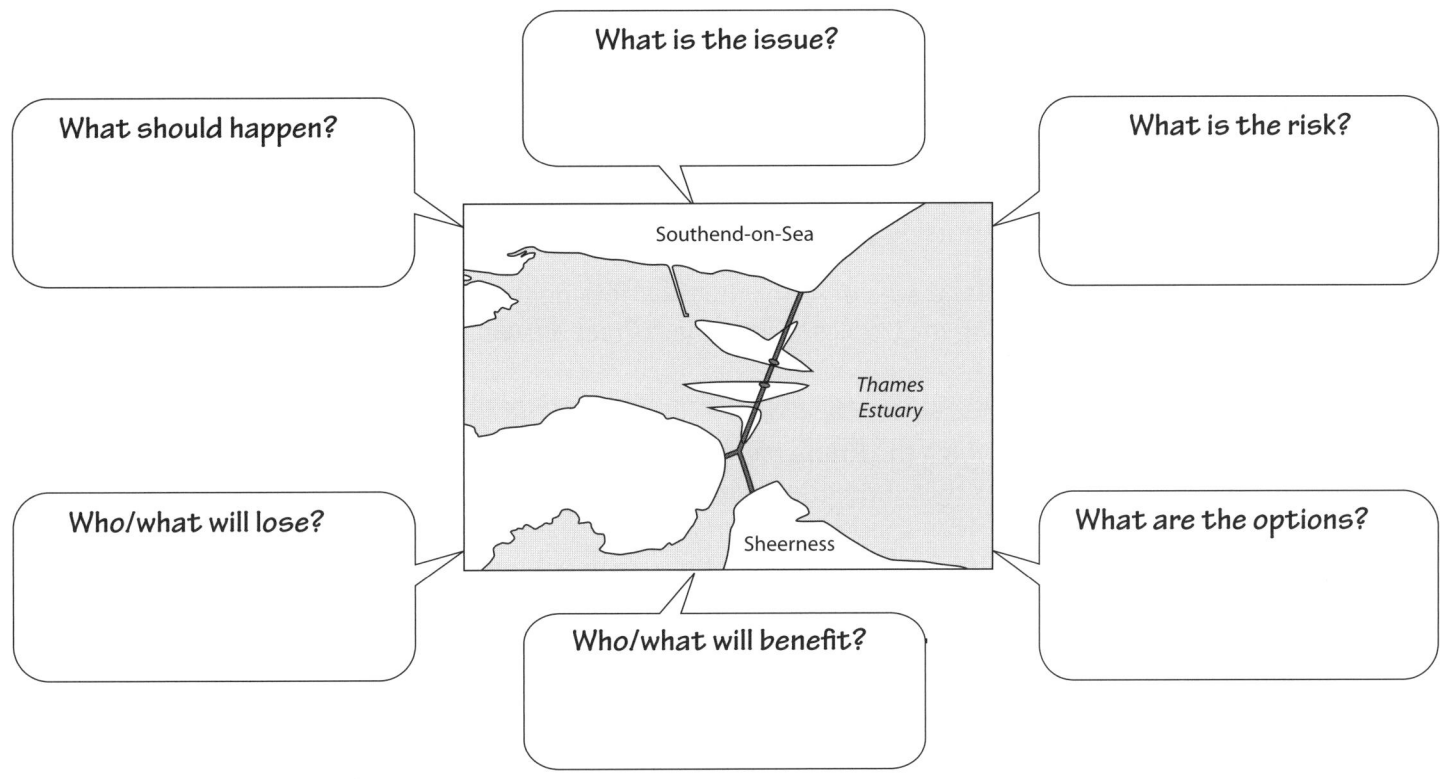

# Africa

## 6.1 What and where is Africa?

**This is about locating the continent of Africa on the world map.**

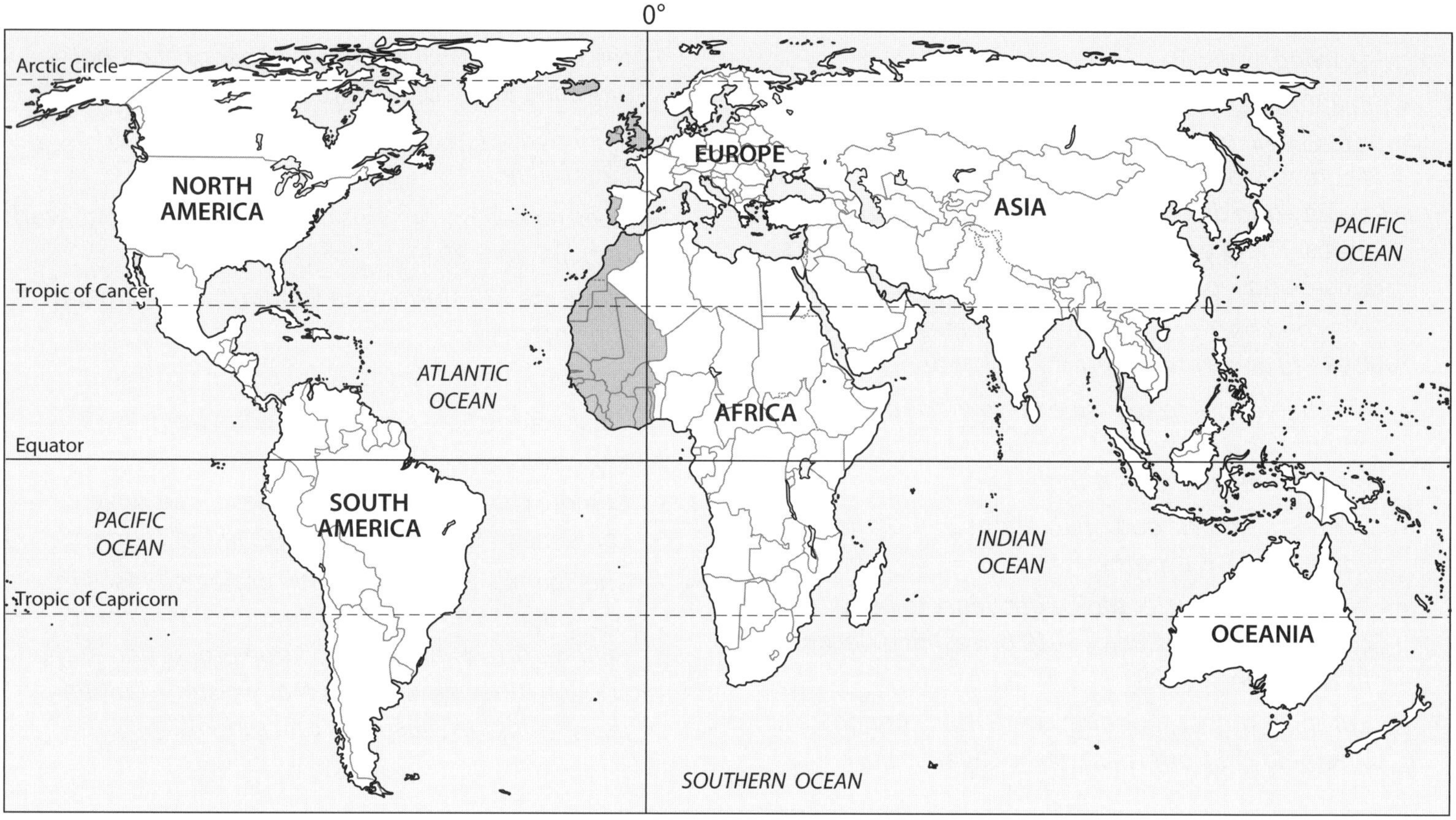

1 Colour in all the land in the world that lies between the Tropic of Cancer and the Tropic of Capricorn. This area is known as 'the tropics' and is the hottest part of the world.

   **a** Which continent has most of its land area inside the tropics? _____

   **b** Which continent lies completely outside the tropics? _____

2 Using a second colour, fill in all the land area of Africa that lies outside the tropics. What proportion of Africa lies outside the tropics? Circle the correct answer.

   About 60%          About 75%          About 35%

3 The map also shows 0° longitude, which the Prime Meridian. It runs through Greenwich in London. The countries that share the Greenwich Meridian Time Zone (GMT) with the UK have been shaded on the map. How many of the countries in Africa which share the same time zone as us can you name?

_____

_____

46 Africa

# 6.2 A little history

**This is an exercise in placing key events in African history into the correct order.**

1 The list of ten key events in African history below has been jumbled up. Put the list into the right order in which the events occurred by writing a number from 1 to 10 in each box.

   About 30 BCE. The Ancient Egyptian civilisation ends. ☐

   About 100 000 years ago. *Homo sapiens* leaves Africa. ☐

   About 1800. The Atlantic Slave Trade is ended. ☐

   1951. Libya is the first African colony to gain independence. ☐

   About 2 million years ago. The first species of human appears. ☐

   About 800 CE. The Mali Empire begins. ☐

   About 300 000 years ago. *Homo sapiens* appears. ☐

   1884. The Berlin Conference divides Africa up between the European nations. ☐

   1420. Portuguese exploration of Africa begins. ☐

   About 3000 BCE. The Ancient Egyptian civilisation begins. ☐

2 Two African countries, shaded on the map below, were never European colonies. Name these two countries and write about them below. Do some research!

   Country A ............................................................

   ............................................................

   ............................................................

   Country B ............................................................

   ............................................................

   ............................................................

Africa 47

# 6.3 What's Africa like today?

**Here you will look at population growth figures and ask questions about them.**

Africa has around 1.3 billion people, and this number is growing fast.
But the figures for the whole of Africa do not tell the full story.

1. The table below shows the population figures (in millions) for six of the largest African countries for 1950 and 2019 and the projected figures up to 2100.

|  | 1950 | 2019 | 2025 | 2050 | 2100 |
|---|---|---|---|---|---|
| Democratic Republic of Congo | 12 | 87 | 104 | 194 | 362 |
| Egypt | 22 | 100 | 112 | 160 | 225 |
| Ethiopia | 18 | 112 | 130 | 205 | 294 |
| Nigeria | 38 | 201 | 233 | 401 | 733 |
| South Africa | 14 | 59 | 63 | 76 | 79 |
| Tanzania | 8 | 58 | 69 | 129 | 286 |

\* UN figures

**a** Between 1950 and 2019, Ethiopia's population increased by 522.2%. Fill in the gaps to complete the calculation to work this out.

1. 112 – _____ = 94
2. _____ ÷ 18 = 5.22
3. 5.22 × _____ = 522.2%

**b** Calculate the percentage increase in Egypt's population between 1950 and 2019. Use the same method as **a**, and show your working.

_____

_____

**Tip!**
To work out the **percentage increase** between two numbers:
1. Calculate the difference between the two numbers you are comparing.
2. Divide the increase by the original number.
3. Multiply the answer by 100 to give a percentage.

You may use a calculator.

**c** Between 1950 and 2019, which country grew:

the fastest? (There are two!) _____ the slowest? _____

**d** Between 1950 and 2050, which country is projected to grow:

the fastest? _____ the slowest? _____

2 **a** What factors can you think of that can cause a large population increase? Try to think of at least three.

_____

_____

**b** The projected figures for South Africa show a small increase in population. What factors might cause a small population increase? Try to think of at least three.

_____

_____

48  Africa

## 6.4 The countries in Africa

This will help you understand Africa's size and the variety of its countries.

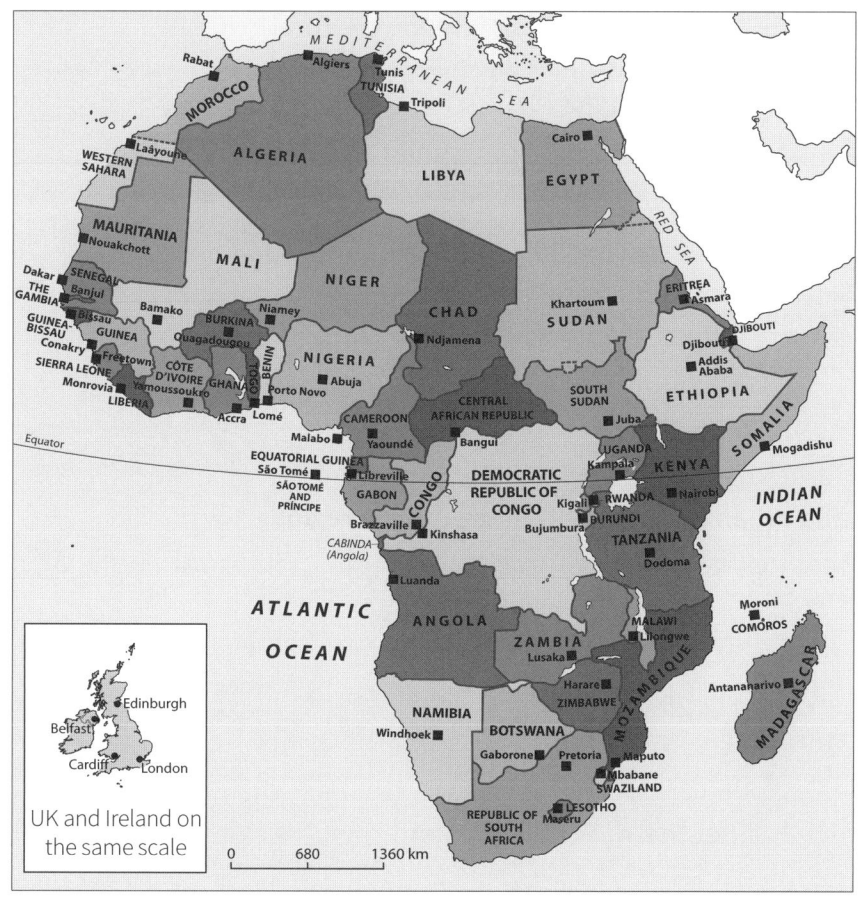

UK and Ireland on the same scale

| Country | |
|---|---|
| Algeria | |
| Angola | |
| Benin | |
| Botswana | |
| Burkina Faso | |
| Burundi | |
| Cameroon | |
| Cape Verde | |
| Central African Republic | |
| Chad | |
| Comoros | |
| Côte d'Ivoire | |
| Dem. Rep. of Congo | |
| Djibouti | |
| Egypt | |
| Equatorial Guinea | |
| Eritrea | |
| Ethiopia | |
| Gabon | |
| Gambia | |
| Ghana | |
| Guinea | |
| Guinea-Bissau | |
| Kenya | |
| Lesotho | |
| Liberia | |
| Libya | |
| Madagascar | |
| Malawi | |
| Mali | |
| Mauritania | |
| Mauritius | |
| Morocco | |
| Mozambique | |
| Namibia | |
| Niger | |
| Nigeria | |
| Republic of Congo | |
| Rwanda | |
| São Tomé and Príncipe | |
| Senegal | |
| Seychelles | |
| Sierra Leone | |
| Somalia | |
| South Africa | |
| South Sudan | |
| Sudan | |
| Swaziland | |
| Tanzania | |
| Togo | |
| Tunisia | |
| Uganda | |
| Zambia | |
| Zimbabwe | |

1  Using the scale on the map above, find out how far it is between the following African capital cities.

   a  Algiers and Pretoria _____    b  Cairo and Abuja _____

   c  Addis Ababa and Bamako _____   d  Mogadishu and Banjul _____

2  Compare distances in the British Isles with distances in Africa. Name two African capital cities that are about the same distance apart as:

   a  London and Cardiff _____   b  Belfast and Cardiff _____

   c  London and Edinburgh _____ .

   d  Into which African country would Great Britain fit easily both north–south and east–west? _____

3  Can you guess the size of Africa's countries? Looking at the map and using a ruler, try to guess what the order of size is of Africa's 54 countries. Write in numbers from 1 (largest) to 54 (smallest) in the boxes alongside the names of the countries. Compare your guesses with a partner's, then look up the answers to see who was the closest!

Note: Cape Verde and Seychelles are not shown on this map. Try looking them up using Google Maps!

Africa

# 6.5 Population distribution in Africa

**This looks at population differences between African regions and selected countries.**

More and more Africans are moving from the countryside to towns or big cities. Here are the rural and urban population percentages for the different African regions.

| Region | Urban | Rural |
|---|---|---|
| Eastern | 25.6% | 74.4% |
| Central | 44% | 56% |
| Northern | 51.6% | 48.4% |
| Southern | 61.6% | 38.4% |
| Western | 45.1% | 54.9% |

*\* UN data*

1 Complete the pie graphs below to show the rural and urban divisions within the African regions. One has been done for you.

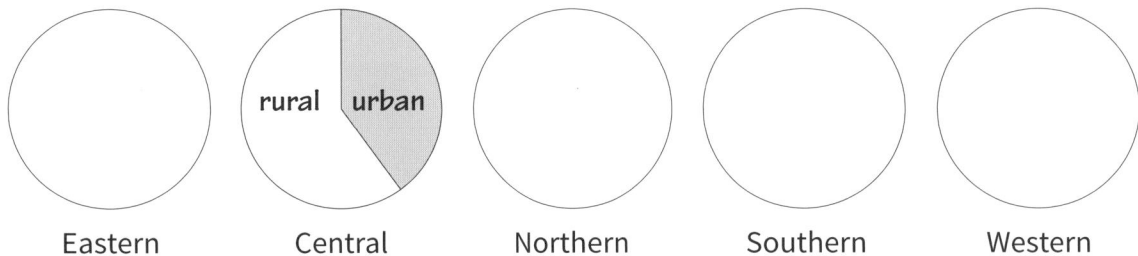

Eastern   Central   Northern   Southern   Western

Look back at the table on page 48 showing population projections for six African countries. They represent all of the African regions.

2 Write down which regions they are in:   Tanzania and Ethiopia ................   Egypt ................

Nigeria ................   South Africa ................   Democratic Republic of Congo ................

3 Suggest reasons why population growth in rural and urban areas might be different.

................................................................................................................................................................

................................................................................................................................................................

Here are the percentages of the populations of the six countries that are under 14 (your generation!).

*\* World Bank data for 2018*

DRC = 46%   Egypt = 33%   Ethiopia = 40%   Nigeria = 44%   South Africa = 29%   Tanzania = 45%

4 Give two possible consequences of having a 'young' population. Explain your answers.

1. ................................................................................................................................................................

2. ................................................................................................................................................................

## 6.6 What are Africa's main physical features?

**Here you will look at the relationship between Africa's physical features and its countries.**

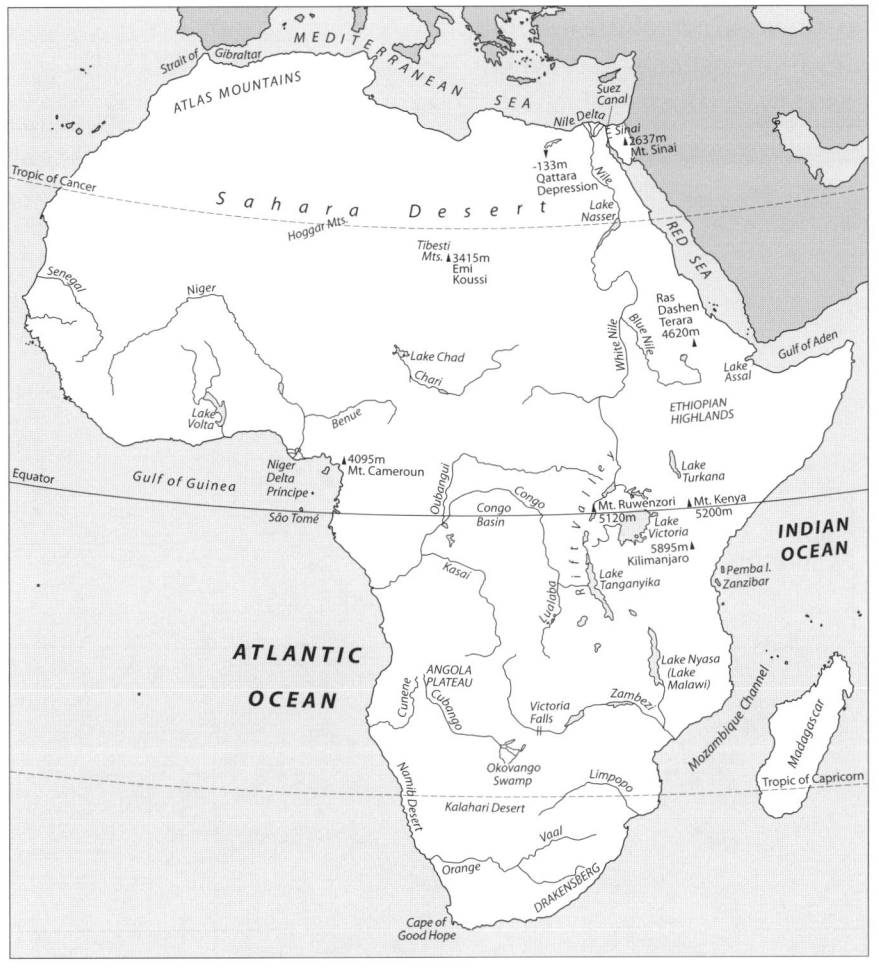

1 Compare this map with the political map on page 49 and answer the following questions:

   a In which countries are the Atlas Mountains?

   b In which countries is the Kalahari Desert?

   c In which country is Mount Kilimanjaro?

   d In which country are the Tibesti Mountains?

   e In which country does the River Benue have its source?

2 On the map above, circle the places where rivers form the borders or part of the borders between African countries. Name these rivers and the countries either side of them. Note that rivers sometimes form part but not all of the border between countries.

3 What is unusual about the Cubango and Chari rivers?

4 Research and find out what is unusual about Lake Volta.

Africa  51

# 6.7 Africa's biomes

**This about telling what biome a place is in from its weather statistics.**

Here are the climate charts for the five cities marked A–E on the map. Using the table, match each one to a numbered chart and write down which of the four biomes it represents. Consider the height above sea level (elevation) of each place when looking at the temperature figures. Also fill in the reasons you chose each chart.

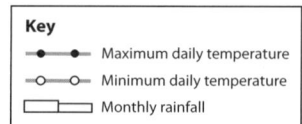

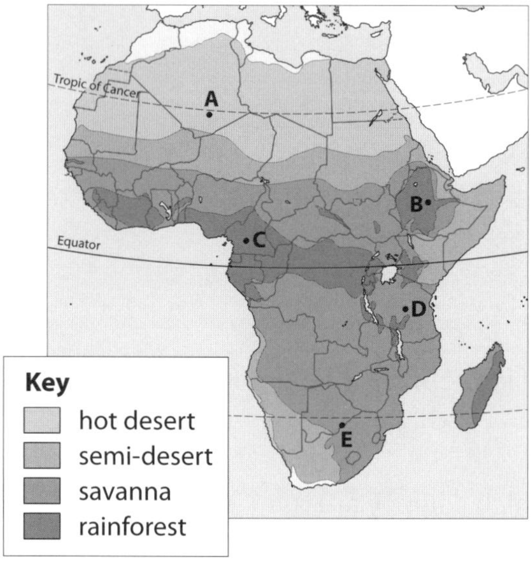

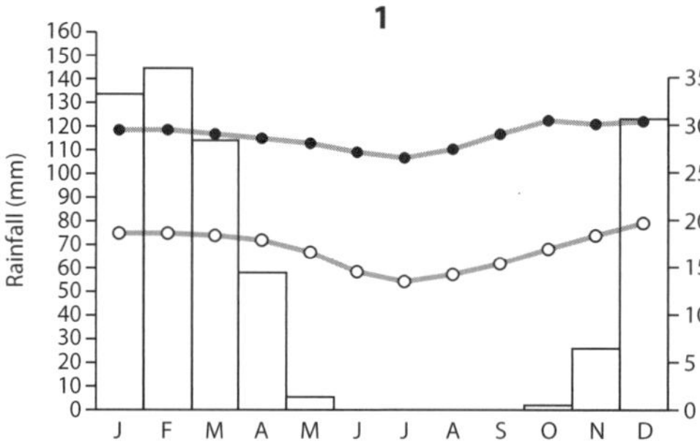

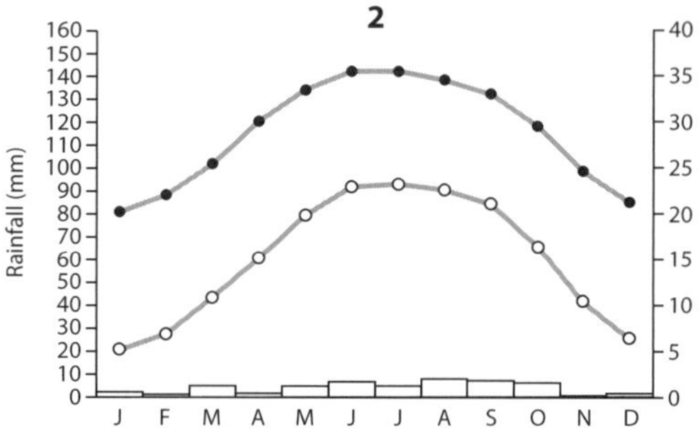

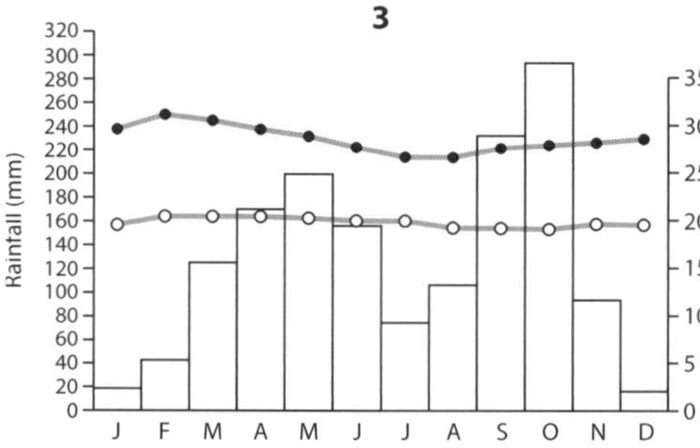

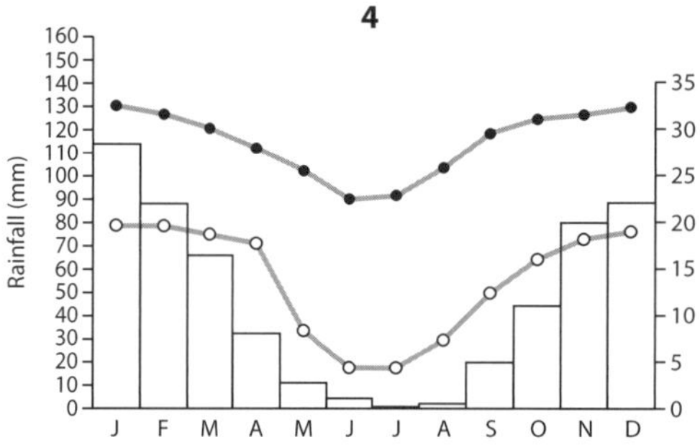

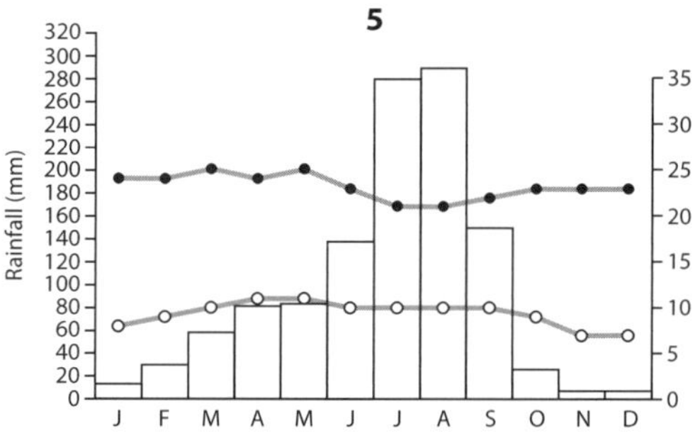

| City | Elevation (m) | Biome | Chart number | Reasons |
|---|---|---|---|---|
| A: Tamanrasset | 1320 | | | |
| B: Addis Ababa | 2355 | | | |
| C: Youndé | 726 | | | |
| D: Dodoma | 1120 | | | |
| E: Gaberone | 983 | | | |

52  Africa

# Kenya

## 7.1 Hello Kenya!

**Here you will look at an overview of Kenya.**

1   a   Kenya is in East Africa. On the map below, find and colour in Kenya.

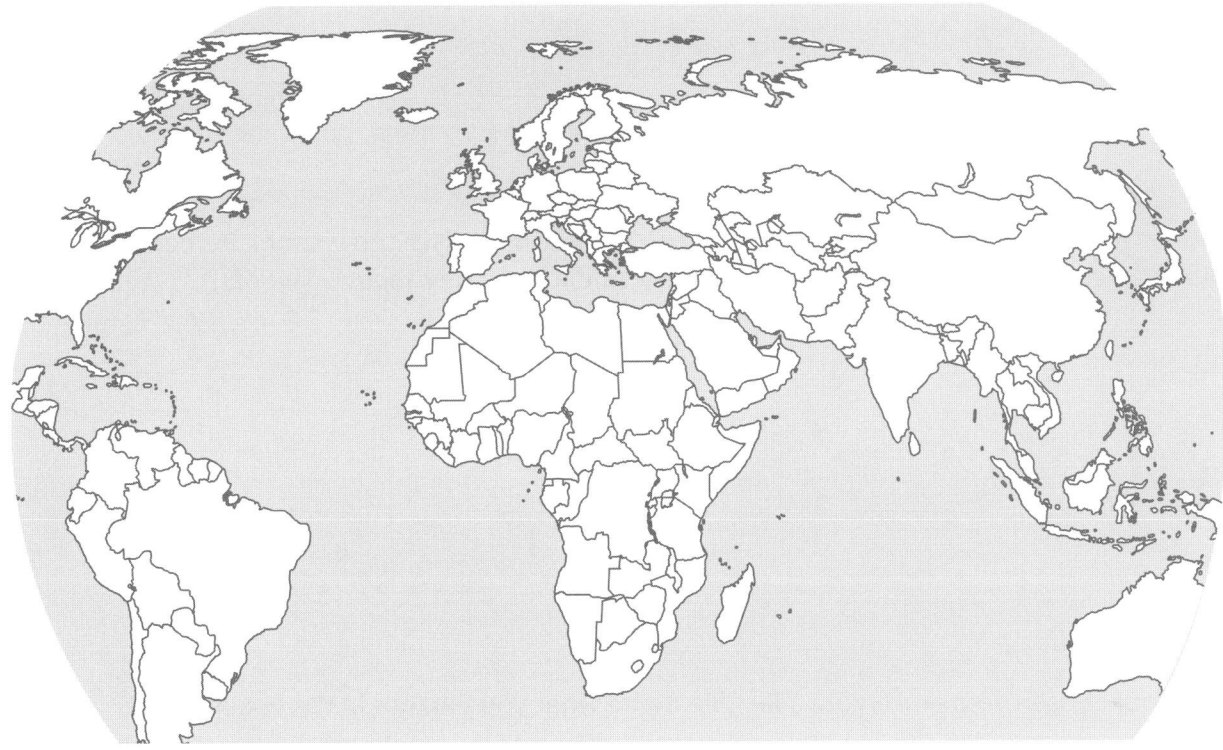

   b   Name the two countries that border Kenya to the north. ........................................

   c   Name the country that borders Kenya to the south. ........................................

   d   Name the ocean to the east of Kenya. ........................................

   e   What is the name of the large lake in the north of Kenya? (Use page 116 of *geog.1* to help you.) ........................................

   f   The Equator runs through Kenya. Draw it on this map using a coloured pencil or a dotted line. (Use page 116 of *geog.1* to help you.)

2   a   How would you get from the UK to Kenya if you were travelling by sea? Draw your route on the map.

   b   Now describe your route in as much detail as possible.

   ........................................................................................................................

   ........................................................................................................................

   ........................................................................................................................

   ........................................................................................................................

**Tip!** Remember to use the names of countries and oceans/seas in your answer. Use an atlas to help you.

# 7.2 What are Kenya's main physical features?

**This is where you will find out more about Kenya's physical features.**

1. Four of Kenya's main physical features are listed below. Which area would you most like to visit? Rank them 1–4, with 1 being your first choice. Give two reasons for your answer.

   Mount Kenya ☐   Lake Turkana ☐   Great Rift Valley ☐   the coast ☐

   Reason 1 ........................................................................................................................

   ........................................................................................................................

   Reason 2 ........................................................................................................................

   ........................................................................................................................

2. Write a fifty word radio advert that would 'sell' your choice to tourists and visitors. Remember your use of adjectives and geographical vocabulary. And it must be *exactly* 50 words!

3. Research your chosen area and create a fact file. Write what you find out in the boxes below.

   Fact 1:

   Fact 2:

   Fact 3:

   Fact 4:

   Fact 5:

   Fact 6:

   My area is ................................

54  Kenya

# 7.3 What's Kenya's climate like?

**This is about Kenya's climate and its climate zones.**

1. Look at the climate facts below. Next to each, write one effect that each would have on farmers.

   | Climate fact | Effect |
   | --- | --- |
   | The lower land is hot all year. | |
   | Rain does not fall steadily through the year. | |
   | Some areas have two rainy seasons each year. | |
   | In many areas, July temperatures exceed 30°C. | |
   | During some years the rains fail completely. | |

2. You have been asked to write a summary of the climate in Kenya for a geography website. Use maps A, B and C on page 120 of *geog.1* to help you.

   **Tip!** Remember to use geographical terms and compass directions in your answer!

3. Imagine you are a Kenyan farmer. Describe where you think it would be best to farm and give reasons for your answer. Write your answer in the speech bubble below.

Kenya 55

# 7.4 A short history of Kenya

**Here you will examine Kenya's history.**

1. Draw your own illustrations for each of these key events in Kenya's history.

| | | |
|---|---|---|
| 5000 years ago: Kenya is empty of humans, except for small groups of hunter-gatherers. | 800 CE: Bantu groups along the coast trade with Arabs from the Middle East. | 1498–1730: Portuguese explorers fight for control of the trade routes, but the Arabs eventually drive them out. |
| 1884: At the Berlin Conference, Britain is given control of the land that is now Kenya. | 1952: The Mau Mau, a group of mostly Kikuyu Kenyans, start a fierce revolt against the British. | 1963: After years of fighting, Kenya gains its independence and elects its first president, Jomo Kenyatta. |

2. What, in your opinion, is the most important difference between Kenya in 1950 and Kenya today? Justify your answer.

......................................................................................................................................................

......................................................................................................................................................

......................................................................................................................................................

## 7.5 Kenya's population

Here you will look at how many people live in Kenya, where they live, and how old they are.

1 Unscramble the terms in brackets to complete these sentences.

When lots of people live in an area, it has a high _____. (pinuooplta ydteins)

In Kenya, most people live in areas that have higher than average _____. (lalaifrn)

2 Using the map below, describe the pattern of population density in Kenya.

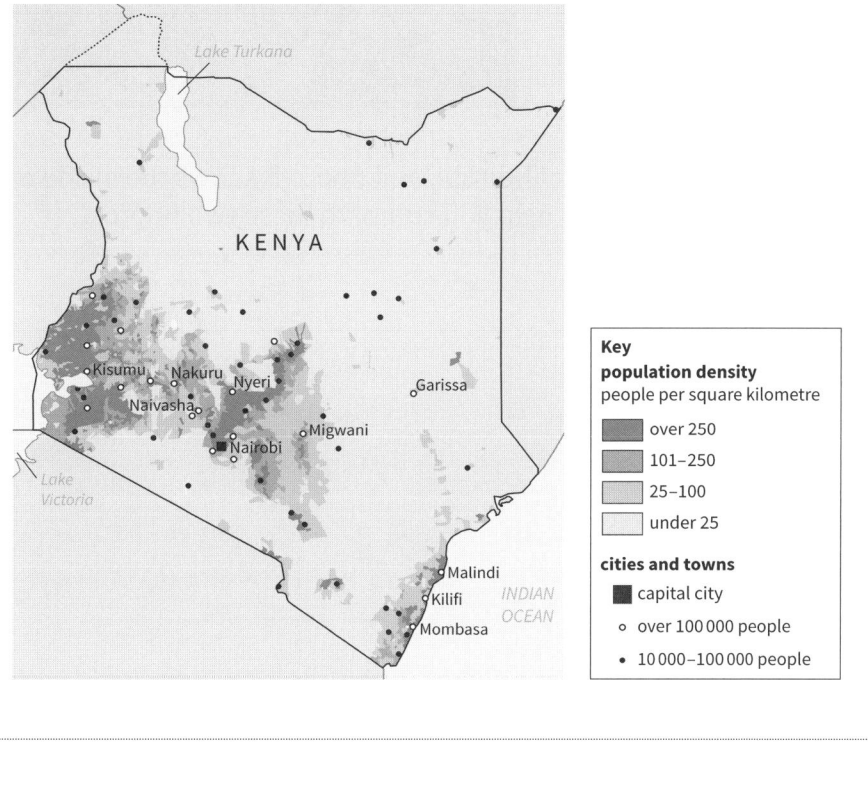

_____

_____

_____

3 Circle the correct word in each pair to complete this paragraph.

Kenya's population is **falling** / **rising**. This is partly due to better **healthcare** / **schools**, meaning that people can receive better treatment for illnesses. Kenya also has quite a **high** / **low** fertility rate. This means that Kenyan women are becoming **less** / **more** likely to have more children. Kenya's population is quite **old** / **young**, which is shown on a population pyramid as a **wide** / **narrow** base.

Kenya 57

# 7.6 What's Nairobi like?

**Here you will look at Kenya's capital city.**

A

B

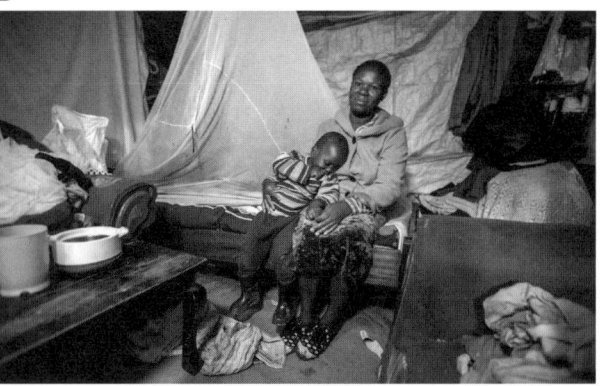

1. Photo A shows a modern house in the west of Nairobi, and photo B shows the inside of a shack in a slum area. Use the table below to help you compare these two photos.

|  | Photo A | Photo B |
|---|---|---|
| Does it look wealthy or poor? How can you tell? |  |  |
| What would it be like to live here? What facilities might there be? |  |  |
| Would you like to live here? Why or why not? |  |  |

2. The infrastructure of a city is the basic structure and features that allow it to function. One example of infrastructure is Nairobi's road network. How many other examples can you think of? Aim for at least five.

   ......................................................................................................................................................

   ......................................................................................................................................................

3. Kibera is one of Nairobi's slums. Its infrastructure is poor. Write down two problems this might cause, and explain why.

   1. ...................................................................................................................................................

   ......................................................................................................................................................

   2. ...................................................................................................................................................

   ......................................................................................................................................................

Kenya

## 7.7 What does everyone do?

**Here you will look at how people make a living in Kenya.**

1 Circle the correct answer to complete these sentences.

   a The most common job in Kenya is **teaching** / **farming** / **fishing**.

   b Pastoralists are farmers who **rear animals** / **grow vegetables** / **build plantations**.

   c Some pastoralists are always on the move: they are **nomads** / **roamers**.

   d Climate change is leading to more frequent **tsunamis** / **droughts** / **hurricanes**, making life harder for the pastoralists.

2 Explain why the Kenyan government might want many more factories.

   ................................................................................................................................

   ................................................................................................................................

3 a Some people work to provide services. What type of industry sector is this?

   (Look at Box C on page 53 of *geog.1* to help you.) ................................

   b How many jobs can you think of in this sector? Aim for at least six.

   ................................................................................................................................

   ................................................................................................................................

   c In the space below, draw an advert for a new mobile phone. Include information about how it can help Kenyans in their daily lives.

**Tip!** Use the information on page 129 of *geog.1* to help you.

Kenya

# 7.8 How Kenya earns money from flowers

**Here you will look at how Kenya exports cut flowers.**

1  In 2005, Kenya exported 5% of the world's total cut flowers. By 2015, it exported 11% and was the third largest exporter of flowers. Suggest what Kenya's flower industry might look like in 2025.

...................................................................................................................................

...................................................................................................................................

...................................................................................................................................

...................................................................................................................................

**Tip!**
Think about the following:
- Do you think the industry will continue to grow?
- If so, why? If not, why not?

Use the information on pages 130–131 of *geog.1* to help you.

2  Read the opinions of Miriam, Joseph and Silas, on page 131 of *geog.1*. Then complete this table about the disadvantages of flower farming in Kenya highlighted by each person.

| Negative impacts: | … for the environment | … for the workers |
|---|---|---|
| Miriam's story | | |
| Joseph's story | | |
| Silas' story | | |

60  Kenya

# 7.9 On safari!

**Here you will look at the impact of the tourist industry in Kenya.**

1  **a**  Read these statements and decide whether they are something the Kenyan government or the local people might say. Colour them in two different colours.

> Tourism brings in money, which helps us build schools and hospitals.

> We've been moved off our land to make way for nature and game reserves.

> Wild animals attack our cattle, but we're not allowed to shoot them because of the reserve.

> We need the reserves to protect the wildlife, which is very important for the tourist industry.

> The tourists are valued more than we are, even though we've been here for longer.

**Key**

**b**  Now draw a key for your two colours in the box.

2  Unscramble the terms in brackets to complete these sentences.

Tourism in Kenya can cause _____. (nciflcot)

One way to improve tourism in Kenya is to make it more _____. (sultaibeans)

An example of this is to limit the _____ of tourists who visit. (mbuern)

Another is to make sure that the _____ are shared fairly with local people. (tpoisrf)

3  In your own opinion, is tourism in Kenya a good thing or a bad thing overall? Justify your answer.

_____
_____
_____
_____

Kenya

# 7.10 So how is Kenya doing?

**Here you will look at how Kenya is doing and how it might do in the future.**

1. Imagine you live in Kenya. You have been asked to give your five greatest hopes for the country in the future – your own wish list of progress or changes that you would like to see happen. What would you put on your wish list and why? Use any data from Chapter 7 to help support your answer.

   **Wish 1** _____

   **Wish 2** _____

   **Wish 3** _____

   **Wish 4** _____

   **Wish 5** _____

2. Now think about everything that you have learned about Kenya. Draw a picture or word map that represents what you understand the country to be like. Use labels and try to be as creative as possible!

62  Kenya

# Your own notes...

# geog.1 workbook

**5th edition**

This workbook is ideal for homework or independent study. It provides:
- engaging activities
- skills development
- support for every double-page spread in the *geog.1* student book

geog.123 is a comprehensive course that matches the National Curriculum at Key Stage 3.

**Did you know?**
- There are glaciers on every continent ...
- ... and in more than 40 countries.

**Did you know?**
- Earth is flatter at the South Pole than the North Pole... because of the weight of Antarctica's ice.

The world's trusted geospatial partner

**Also available in the series**

Kerboodle provides online Lessons, Resources and Assessment to support teaching and learning in the classroom and at home. Online versions of the student book are also available.

geog.1 workbook – pack of 10
ISBN 978 0 19 844605 7

geog.1 workbook answer book
ISBN 978 0 19 844607 1

Printed on paper produced from sustainable forests.

**OXFORD**
UNIVERSITY PRESS

**How to get in touch:**
web   www.oxfordsecondary.co.uk
email  schools.enquiries.uk@oup.com
tel    +44 (0)1536 452620
fax   +44 (0)1865 313472

ISBN 978-0-19-844606-4